The Manufacture of Methyl Ethyl Ketone from 2-Butanol

A WORKED SOLUTION TO A PROBLEM IN CHEMICAL ENGINEERING DESIGN

The Manufacture of Methyl Ethyl Ketone from 2-Butanol

A WORKED SOLUTION TO A PROBLEM IN
CHEMICAL ENGINEERING DESIGN

by

D. G. Austin   and   G. V. Jeffreys

Chemical Engineering Department
University of Aston in Birmingham

THE INSTITUTION OF CHEMICAL ENGINEERS
in association with
GEORGE GODWIN LIMITED

Published by
The Institution of Chemical Engineers,
George E. Davis Building,
165–171 Railway Terrace,
Rugby,
Warwickshire,
CV21 3HQ

ISBN 0 85295 115 9

in association with
George Godwin Ltd.,
1–3 Pemberton Row,
Red Lion Court,
Fleet Street,
London
EC4P 4HL

ISBN 0 7114 4616 4

© The Institution of Chemical Engineers 1979

All rights reserved. No part of this publication may be reproduced, stored in a retrieval system, or transmitted, in any form or by any means, electronic, mechanical, photocopying, recording, or otherwise, without the prior permission of the copyright owner.

Printed by Warwick Printing Co Ltd, Theatre Street, Warwick.

CONTENTS

			Page
FOREWORD *by R. Parkins*			(vii)
ACKNOWLEDGEMENTS			(viii)
CHAPTER 1	—	Introduction	1
CHAPTER 2	—	The Problem	3
CHAPTER 3	—	General Considerations	7
CHAPTER 4	—	Process Design — Reactor	15
CHAPTER 5	—	The Condenser	65
CHAPTER 6	—	The Absorption Column	81
CHAPTER 7	—	The Solvent Extraction Unit	93
CHAPTER 8	—	The Product Distillation Unit	118
CHAPTER 9	—	The Heat Balance	135
CHAPTER 10	—	Mechanical Design of Reactor	138
CHAPTER 11	—	Economic Evaluation of the Process	149
CHAPTER 12	—	Operability Study of the Reactor Section	167
CHAPTER 13	—	Instrumentation and Control	179
CHAPTER 14	—	Storage of Process Materials	185
CHAPTER 15	—	Final Considerations	187
REFERENCES			188
APPENDICES			
	A	— Vapour Heat Capacities	190
	B	— Heat of Reaction	195
	C	— Effective Thermal Conductivity of Packed Bed	196
	D	— Reactor Design Logic Flow Diagram	198
	E	— Solution of Differential Equations for Reactor Design	206
	F	— Physical Properties of Reactor Fluids	207
	G	— Specific Heat of Flue Gas	211
	H	— Data for Reboiler Design	211
	I	— Vapour Thermal Conductivities	214
	J	— Physical Properties of the Condensate	215
	K	— Economic Analysis Logic Flow Diagram	217
APPENDIX REFERENCES			221

FOREWORD

This book has been prepared to demonstrate how chemical engineering principles can be applied to the design of process equipment, by reference to the Design Project examination question set in 1974. The object is to demonstrate the application of principles, rather than to design a methyl ethyl ketone process in every detail.

Although written primarily to meet the needs of candidates preparing for the Design Project examination of the Institution of Chemical Engineers, the work, being concerned with the practical application of basic principles, will be of value to all chemical engineers.

For that reason we welcome the commercial publication of a hardback edition in addition to the student paperback published by the Institution. This should be a valuable reference in all chemical engineering libraries and design offices.

The text illustrates how solutions to the design problems involved can be derived by reference to standard chemical engineering texts and other reference books generally available. It is believed that the computer programs included are also generally available and that the numerical solutions given could be obtained with the aid of a pocket calculator.

The Design Project examination of the Institution of Chemical Engineers is intended to test the candidate's knowledge of chemical engineering principles and his understanding of the inter-relationship of the various disciplines already covered in Parts 1 and 2 of the CEI examinations. The principal requirement is that he should demonstrate an adequate knowledge of these disciplines by applying them to a real problem of industrial significance.

Since the last worked solution to a Design Project examination question was published in 1961 there have been many developments in chemical engineering design. Thus, in 1961 design calculations were performed manually, whereas nowadays computers are generally available so that additional parameters can be included in the design calculations. Furthermore there is now a greater emphasis on safety, and in consequence chemical engineers are required to make operability studies at the design stage. The Institution of Chemical Engineers is therefore obliged to test a candidate's knowledge in these subjects.

In working through the Design Project, candidates may not be able to cover all items to the degree of thoroughness illustrated in this text. For each item of design work attempted however, candidates should aim for the standard indicated. Reference is made to previous publications prepared with similar objectives in view, so as to lead into the present edition which is intended to give further assistance in preparing for the examination in its present form.

In commending the book to candidates preparing for the Institution's Design Project examination, I would also on behalf of the Board, like to thank Dr D.G. Austin and Professor G.V. Jeffreys for once again willingly accepting the responsibility for producing it.

R. PARKINS
Chairman – Board of Examiners

ACKNOWLEDGEMENTS

The authors would like to express their thanks to members of staff of the Chemical Engineering Department of Aston University for their helpful discussions during the preparation of this text; particularly Dr A.V. Bridgwater for his advice and generous assistance in the preparation of the economic analysis in Chapter 11 and to Dr D.A. Lihou for preparing and writing the Operability Analysis of the reactors in Chapter 12. In addition our thanks are given to Mr C.L. Hung, a postgraduate student in the Department for preparing the Appendix on the correlation of specific heats. Finally, we are most grateful to Miss Joy Harris for typing the manuscript, photocopying the typescript and generally organising the text for publication.

D.G. Austin and G.V. Jeffreys
Department of Chemical Engineering
University of Aston in Birmingham

Chapter 1

INTRODUCTION

1.1 Background information

Since 1972 the "Design Project" is the only examination set by the Institution of Chemical Engineers, and is open to candidates who have successfully completed Parts 1 and 2 of the Council of Engineering Institutions examinations, or have obtained exemption from the CEI examinations. This design project test replaces the former "Part 3" in name only and the definition of the exercise is the same as that formulated in 1958. That is, the project is always set in such a way that the design can be performed and the report written with the assistance of the standard chemical engineering texts, the *Chemical Engineers' Handbook* and other handbooks together with the well known journals normally available in most factory, college and public libraries. The examiners of the Institution recognise that programmable calculators and computers are now widely available and encourage their use provided that the programs were written by the candidates or are generally available in recognised technical publications. In addition the Institution expects the complete answer to be prepared from first principles.

In the following example of a design report it has been necessary to discuss in detail certain theoretical points to a far greater extent than would usually be required in an actual report of an intended project. This has been done only when reference cannot be made to one of the excellent chemical engineering texts available. Thus it is assumed that all candidates about to undertake the design project examination would possess the following books as a basis of their own personal technical library.

> *Chemical Engineering,* Vols 1, 2 & 3.
> Pergamon Press, London 1978
> Coulson, J.M. & Richardson, J.F.
>
> *Mass Transfer Operations*
> McGraw Hill, New York, 1972
> Treybal, R.E.
>
> *Chemical Reaction Engineering*
> Wiley, New York, 1973
> Levenspiel, O.V.
>
> *Absorption & Extraction*
> McGraw Hill, New York, 1976
> Sherwood, T.K. & Pigford, R.E.

Mathematical Methods in Chemical Engineering
Academic Press, 1978
Jenson, V.G. & Jeffreys, G.V.

Finally, it is still worth quoting from the original *Problem in Chemical Engineering Design* by J.M. Coulson and Sir Frederick Warner: *"The design of any chemical installation requires the co-operation of the chemist, chemical engineer and mechanical engineer. Each stage of the development of any design is the result of proposals which have been examined from the standpoint of what is efficient and what is practical. With this co-operation a design is finally arrived at which can be passed to the drawing office for detailing and layout. In practice this initially involves the preparation of flowsheets showing the quantities involved and indicating alternative routes. This is accomplished by the assembly of the necessary physical data which will allow the making of energy balances".*

In some cases a number of calculations have to be made where several operating conditions may be varied. For example in the design of extraction equipment the solvent/feed ratio affects all the design calculations and these must be made repeatedly before an energy balance can be prepared and the equipment specified. Nowadays this is conveniently achieved by computer but it has tended to induce the design engineer to include more variables into his calculations so that the final design is more precise. An example of this is given in the design of the catalytic reactor where both longitudinal and lateral temperature profiles have been considered and the rate of decay of the catalyst activity has been included in the calculations.

Chapter 2

THE PROBLEM

2.1 General

The problem considered here is that set for the 1974 examination and is given in the form in which it was received by the candidates.

The Institution of Chemical Engineers

The Design Project

1st October to 1st December 1974

Instructions for the Design Project, 1974

BEFORE STARTING WORK READ carefully the enclosed copy of *The Regulations for The Design Project* in conjunction with the following details for The Design Project for 1974.

In particular, candidates should note that all the questions should be answered in the section headed "Scope of Design Work Required".

The answers to the Design Project should be returned to The Institution of Chemical Engineers, 15 Belgrave Square, London SW1X 8PT, by 1700 hours on December 1st 1974. In the case of overseas candidates, evidence of posting to the Institution on November 30th will satisfy this requirement. The wrappings must be marked on the OUTSIDE with the Candidate's name and the words: "DESIGN PROJECT".

The Design Project will be treated as a test of the ability of the candidate to tackle a practical problem in the same way as might be expected if he were required to report as a chemical engineer on a new manufacturing proposal. The answers to the Design Project should be derived by the application of fundamental principles to available published data; they should on no account include confidential details of plant or processes which may have been entrusted to the candidate. Particular credit will be given to concise answers.

References must be given of all sources of published information actually consulted by the candidate.

The answers should be submitted on either A4 or foolscap paper but preferably on A4. Squared paper and drawing paper of convenient size may be used for graphs and drawings respectively. The text may be handwritten or, preferably, typewritten; in the latter case it is permissible for another person to type the final copies of the answers. Original drawings should be submitted. Copies, in any form, will not be accepted.

Each sheet and drawing must be signed by the candidate and this signature will be taken to indicate that the sheet or drawing is the candidate's unaided work, except typing. In addition, the declaration forms enclosed must be filled in, signed, witnessed and returned with the answers. The manuscript, drawings, and any other documents should be fastened in the folder supplied, in accordance with the instructions appearing thereon.

Answers to the Design Project itself must be written in the English language and should not exceed 20 000 words excluding calculations.

The use of SI units is compulsory.

Candidates may freely utilise modern computational aids. However, when these aids are employed, the candidate should clearly indicate the extent of his own contribution, and the extent of the assistance obtained from other sources. For computer programs which have been prepared by the candidate himself, a specimen print-out should be appended to the report. Programs from other sources should only be used by the candidate provided adequate documentation of the program is freely and publicly available in recognised technical publications. The candidate must demonstrate clearly that he fully understands the derivation of the program, and the significance and limitation of the predictions.

The answers submitted become the property of the Institution and will not be returned in any circumstances.

1974 DESIGN PROJECT

The Project

Design a plant to produce 1×10^7 kg/year of methyl ethyl ketone (MEK).

Feedstock:— Secondary butyl alcohol.

Services available:—
 Dry saturated steam at 140°C
 Cooling water at 24°C
 Electricity at 440 V 3-phase 50 Hz
 Flue gases at 540°C.

The process

The butyl alcohol is pumped from storage to a steam-heated preheater and then to a vaporiser heated by the reaction products. The vapour leaving the vaporiser is heated to its reaction temperature by flue gases which have previously been used as reactor heating medium. The superheated butyl alcohol is fed to the reaction system at 400°C to 500°C where 90% is converted on a zinc oxide — brass catalyst to methyl ethyl ketone, hydrogen and other reaction products. The reaction products may be treated in one of the following ways:—

(a) Cool and condense the MEK in the reaction products and use the exhaust gases as a furnace fuel.

(b) Cool the reaction products to a suitable temperature and separate the MEK by absorption in aqueous ethanol. The hydrogen off gas is dried and used as a furnace fuel. The liquors leaving the absorbers are passed to a solvent extraction column, where the MEK is recovered using trichlorethane. The raffinate from this column is returned to the absorber and the extract is passed to a distillation unit where the MEK is recovered. The trichlorethane is recycled to the extraction plant.

Scope of Design Work Required
All questions must be answered

1. Prepare material balances for the two processes.
2. On the basis of the cost data supplied below decide which is the preferable process.
3. Prepare a material flow diagram of the preferred process.
4. Prepare a heat balance diagram of the preheater—vaporiser—superheater—reactor system.
5. Prepare a chemical engineering design of the preheater—vaporiser—superheater—reactor system and indicate the type of instrumentation required.
6. Prepare a mechanical design of the butyl alcohol vaporiser and make a dimensioned sketch suitable for submission to a drawing office.

Process data

Outlet condenser temperature = 32°C.
Vapour and liquid are in equilibrium at the condenser outlet.
Calorific value of MEK = 41 800 kJ/kg.

Cost data

Selling price of MEK	=	£ 9.60 per 100 kg
Steam raising cost	=	£ 0.53 per 10^6 kJ
Cost of tower shell	=	£2 000
Cost of plates	=	£2 000
Cost of reboiler	=	£2 500
Cost of heat exchanger (per distillation column)	=	£8 000
Cost of solvent extraction auxiliaries	=	£1 000
Cost of absorbtion and distillation column packing, supports and distributors	=	£2 000
Cost of tanks (surge, etc)	=	£1 000
Cost of control of whole plant	=	£9 000
Cost of instrumentation for control of recovery section	=	£4 500
Cost of electricity for pumps	=	£5 000
Pump costs (total)	=	£3 000
Cost of cooling water for whole plant	=	£5 000

Reactor data

The "short cut" method proposed in Ref.1 may be used only to obtain a preliminary estimate of the height of catalyst required in the reactor. The reactor should be designed from first principles using the rate equation, below, taken from Ref.1.

$$r_A = \frac{C(P_{A,i} - P_{K,i} P_{H,i}/K)}{P_{Ki}(1 + K_A P_{A,i} + K_{AK} P_{A,i}/P_{K,i})}$$

where $P_{A,i}$, $P_{H,i}$, and $P_{K,i}$ are the interfacial partial pressures of the alcohol, hydrogen and ketone in bars, and the remaining quantities are as specified by the semi-empirical equations below:—

$$\log_{10} C = -\frac{5964}{T_i} + 8.464$$

$$\log_{10} K_A = -\frac{3425}{T_i} + 5.231$$

$$\log_{10} K_{AK} = +\frac{486}{T_i} - 0.1968$$

In these equations, the interfacial temperature T_i is in Kelvin, the constant C is in kmol/m² h, K_A is in bar⁻¹, and K_{AK} is dimensionless.

The equilibrium constant, K, is given in Ref.1 (although the original source is Ref.2) by the equation:—

$$\log_{10} K = -\frac{2790}{T_i} + 1.510 \log_{10} T_i + 1.871$$

where K is in bar.

Useful general information will be found in Ref.3.

References

1. Perona, J. J. and Thodos, G. *A. I. Ch. E. Jl*, 1957, 3, 230.
2. Kolb, H. J. and Burwell, R. L. (jr) *J. Amer. chem. Soc.*, 1945, 67, 1084.
3. Rudd, D. F., and Watson, C. C. *"Strategy of Process Engineering"*, 1968. (New York: John Wiley & Sons Inc.).

2.2 Introduction to the problem

Methyl Ethyl Ketone (MEK), or 2-butanone is an important commercial chemical that is produced in large quantities for a wide variety of processes and products. Chemically it is the next higher homologue of acetone that in addition to displaying the characteristic reactions of ketones, undergoes a number of special reactions, such as condensation with aldehydes to form high molecular weight ketones, cyclic compounds and ketals. It is a powerful solvent, only partially miscible in water, has a lower vapour pressure and correspondingly higher boiling point than acetone and these properties are exploited in surface coating, cellulose manufacture and in acrylic resin and vinyl polymer production. In fact nearly all synthetic and natural resins commonly employed in lacquers are soluble in MEK and furthermore it is employed as a solvent in many extraction processes in the chemical and petroleum industries. The annual production rate in 1977 was 100000 tonnes worldwide[1].

There are three basic methods for the manufacture of MEK. These are

(i) Synthesis from refinery gases
(ii) Dehydrogenation from 2-butanol (secondary butyl alcohol)
(iii) Selective oxidation of 2-butanol

Of the different processes it has been found that dehydrogenation of 2-butanol is the most economical and this process has been chosen for design in the following text.

2.3 Summary of the project

The process for the manufacture of MEK consists of dehydrogenating 2-butanol continuously in a reactor containing a composite catalyst of brass maintained at 400°C to 500°C at atmospheric pressure. 90.0% of the butyl alcohol can be converted to MEK and the feasibility of separating the MEK from the reaction products and using the exhaust gases as a furnace fuel, or separating the MEK, recovering and recycling the unconverted alcohol, drying the hydrogen and using this as furnace fuel must be considered. Both of these proposals have been analysed and the results are presented in Chapter 11.

It will suffice to state here that separation of the unchanged alcohol for recycle is the most economical and the design of the recovery system has been undertaken in this design report. Therefore the products of reaction are cooled and condensed in a water-cooled condenser, and the condensate is passed to a distillation unit where the MEK is obtained as distillate and butyl alcohol as the bottom product is recycled and mixed with the reactor feed for reprocessing. The gaseous effluent from the condenser is absorbed in water and the off-gas from the absorber is dried and pumped into the plant fuel system. The liquid discharged from the absorber is treated with trichlorethane to extract the MEK and alcohol and the extract from this column is fractionated to recover the MEK as product and the small amount of alcohol is recycled. The proposed scheme for the continuous production of 1.0×10^7 kg per annum of methyl ethyl ketone (MEK) from secondary butyl alcohol is presented in Figure 2.1, drawn to comply with BS 1553 Part 1, but extended to include the symbolic representation described in *Drawing Office Guide to Symbols used in the Chemical, Petroleum and Allied Industries*[2].

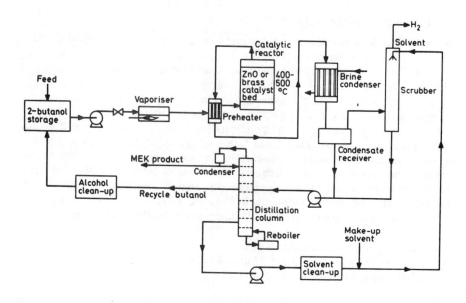

Figure 2.1 — *Production of methyl ethyl ketone from 2-butanol.*

Chapter 3
GENERAL CONSIDERATIONS

3.1 General considerations of the problem

The catalytic dehydrogenation of 2-butanol has been extensively studied by Thodos and his co-workers[3,4] and by Ford and Perlmutter[5]. All these investigations have identified ten possible reaction mechanisms, and all the experimental results indicate that one or more of the following steps are rate controlling.

(i) Adsorption of alcohol vapour on a single active site.
(ii) Decomposition of the alcohol-active site complex to form MEK vapour leaving molecular hydrogen adsorbed on the catalyst.
(iii) Desorption of the molecular hydrogen.
(iv) The adsorbed alcohol molecule reacts with an adjacent vacant active site to produce adsorbed ketone and molecularly adsorbed hydrogen.

Perona & Thodos found that the desorption of hydrogen was the rate-controlling step in the temperature range 350°C to 750°C whereas Thaller & Thodos and Ford & Perlmutter favoured step (iv) over the temperature range 350°C to 400°C. Both the latter research studies confirmed that the rate-controlling step was temperature dependent and it appears that step (iii) becomes rate-controlling outside the close range 350°C to 400°C. Thaller & Thodos' study was undertaken with a differential reactor and Ford & Perlmutter used a stirred vessel reactor whereas Perona & Thodos employed an integral reactor and extrapolated their results to obtain their rate equation. These latter results must therefore be considered to be average results. However, the agreement between their calculated and experimental results is very good and suggests that their rate equation adequately characterises the conversion expected in a commercial scale reactor. Consequently the following rate equation will be utilised for the design of the reactor.

$$r_c = \frac{C_i[p_{Ai} - (p_{Ki}p_{Hi}/K_i)]}{p_{Ki}[1 + K_{Ai}p_{Ai} + K_{AKi}(p_{Ai}/p_{Ki})]} \tag{3.1}$$

where the constants are related to the temperature by the following relations

$$\log C_i = -(5964/T_i) + 8.464 \tag{3.2}$$

$$\log K_{Ai} = -(3422/T_i) + 5.326 \tag{3.3}$$

$$\log K_{AKi} = (269.2/T_i) - 0.1959 \tag{3.4}$$

All the research studies on this reaction report that the butyl alcohol undergoes thermal cracking and that the rate of the cracking reaction was found to increase sharply with temperature above 400°C. However the rate of the dehydrogenation reaction is also temperature dependent and the rate of formation of MEK is negligible below 300°C. Therefore, since this reaction is endothermic, the centre of the reaction tube will have the minimum temperature and this must not be less than 300°C. For the same reason a small amount of cracking must be tolerated and furthermore Perona & Thodos have also shown that the catalyst is regenerated fairly easily so that the upper limit catalyst temperature is 500°C. These two temperatures set the limits for the reactor tube radius.

The products of reaction discharged from the reactor consists of MEK vapour, hydrogen and unchanged alcohol. All these products are reasonably stable below 400°C and it is unnecessary to instal a rapid quench unit. However a considerable amount of heat will be discharged from the reactor which must be recovered by heat exchange with the incoming feed before the MEK and alcohol are condensed.

The majority of the MEK and alcohol discharged from the reactor will be condensed, but since the reaction products contain hydrogen a small amount of MEK and alcohol will leave the condenser saturated in the hydrogen. These vapours will be stripped from the hydrogen by absorption in water, and since the solubility of MEK in water is of the order of 27%, the concentration of the effluent from the absorption should not exceed 27% MEK. The MEK and alcohol in the liquid effluent from the absorber will be recovered by solvent extraction.

The choice of solvent for the recovery of the MEK from the aqueous solution depends on a number of factors which are discussed in depth in Treybal's *Liquid Extraction*.[6] and will not be re-iterated here. Treybal, Newman & Hayworth[7] cite a number of possible solvents for the extraction of MEK from water including trichlorethylene, trichlorethane, hexanes, heptanes and other hydrocarbons. However, from the point of view of ease of recovery of the MEK from the extract solution, it appears that 1:1:2 trichlorethane is the preferred solvent and the equilibrium data has been published by Treybal, Newman and Hayworth[7]. The data

Table 3.1 — *Phase equilibria and density of the system methyl ethyl ketone/ water/1.1.2 trichlorethane.*

MEK % wt	Water % wt	1.1.2 Trichlorethane % wt	Density kg/m^3	MEK % wt	Water %wt	1.1.2 Trichlorethane % wt	Density kg/m^3
18.15	81.74	0.11	973.3	75.00	5.08	19.92	890.4
12.78	87.06	0.16	980.4	58.62	2.73	38.65	972.0
9.23	90.54	0.23	985.3	44.38	1.48	54.14	1055.3
6.00	93.70	0.30	990.2	31.20	1.00	67.80	1142.5
2.83	96.80	0.37	994.2	16.90	0.52	82.58	1255.4
1.02	98.57	0.41	996.6	5.58	0.00	94.42	1362.6

are presented in Table 3.1 and Figure 3.1 together with the density of each phase. There it will be seen that an isopicnic exists between the raffinate containing 11.0% MEK and the extract containing 53% MEK. This could cause flooding or phase inversion in the extraction column which must be avoided by ensuring that the

heated by flue gas. The superheaters will be designed to raise the temperature of the alcohol vapour to 500°C, at which temperature the vapours enter the reactor.

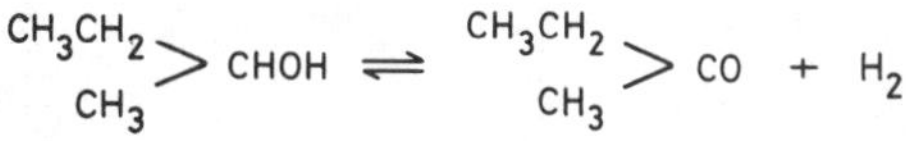

$$\begin{array}{c} CH_3CH_2 \\ CH_3 \end{array} \!\!> CHOH \rightleftharpoons \begin{array}{c} CH_3CH_2 \\ CH_3 \end{array} \!\!> CO + H_2$$

Figure 3.2 — *Dehydrogenation of 2-butanol.*

The 2-butanol is dehydrogenated according to the reaction shown in Figure 3.2 which is reversible, and for which the equilibrium constant is expressed as a function of temperature by[9]

$$\log_{10} K_i = (2790/T_i) + 1.510 \log_{10} T_i + 1.871 \qquad (3.5)$$

The dehydrogenation reaction is endothermic and, as shown in Appendix B, the heat of reaction is 73900 kJ/kg mole at the mean reaction temperature. Therefore a considerable amount of heat will have to be supplied and it is proposed to design a multi-tube reactor heated by flue gas.

3.2.1 *Reactor tube diameter*

Perona & Thodos[3] employed a single tube reactor 1.25 cm in diameter packed with brass beads 0.32 cm in diameter. Although brass spheres are acceptable for an investigation on pilot scale equipment, the cost of fabricating catalyst particles of this shape is prohibitive for the quantity required for three commercial scale reactors. However, it was essential to select a particle shape that closely resembled spheres to ensure that the rate equations developed by Perona & Thodos[3] remain applicable. Therefore right circular cylinders were chosen (0.32 cm diameter, 0.32 cm length) because they have the same specific surface voidage fraction[10] and exhibit similar pressure drop characteristics to those of spheres[11]. Fabrication costs are relatively low since the particles would be manufactured by the cutting of extruded brass rod in one operation. In this experimental reactor some channelling may occur since the ratio (tube diameter/catalyst particle diameter) is only 4:1. However these authors conclude their publication with an example calculation in which they specify a tube diameter of 7.5 cm with catalyst particles 0.32 cm in size. Hence it is necessary to estimate a suitable reactor tube diameter for the commercial scale reactor.

It has been stated above that the reaction is endothermic so that the tube axis temperature will be the minimum and it was shown in Section 3.1 that the reaction rate was very slow below 300°C. Furthermore cracking becomes significant above 500°C and therefore it is necessary to control the reaction between these temperatures. The relationship between heat of reaction, flow rate, tube diameter and temperature can be expressed by the equation [7,8]

$$\frac{\partial^2 T}{\partial x^2} + \frac{1}{x}\frac{\partial T}{\partial x} - \frac{GC_p}{k_E}\frac{\partial T}{\partial z} - \frac{S\Delta H r_c}{k_E} = 0 \qquad (3.6)$$

and the reactor tube diameter may be estimated from equation (3.6). This has been done in Section 4.3 where it will be seen that a suitable diameter would be 4.2 cm and this has been chosen as the basis for the design of the reactor. Consequently, 18:8 stainless steel tubes 4.2 cm nominal bore will be recommended.

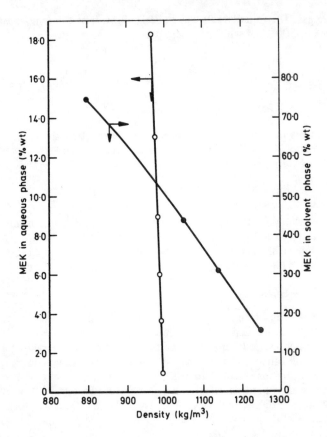

Figure 3.1 — *Location of isopicnic composition.*

concentration of MEK in the effluent from the absorption column fed to the extraction unit does not exceed 10% on a weight basis. The MEK in the extract phase will be recovered by distillation and the regenerated solvent will be recycled to the extraction unit.

The condensate from the condenser will be mixed with the distillate from the solvent recovery column to form the feed to the MEK product distillation unit, and the distillate from this unit will be the MEK product from the process.

3.2 Proposed method of manufacture

The proposed method of manufacture is that summarised in the statement of the problem and will include the general considerations referred to in Section 3 above. That is, the cold feed of 2-butanol will be pumped from the feed tank through a steam heater to a vertical thermo-syphon reboiler in which the alcohol vaporised. The thermo-syphon reboiler will be heated by the reaction produ discharged from the reactor and the wet alcohol vapour will be passed t knock-out drum to remove any entrained liquid. The liquid separated wil recycled and the dry alcohol vapour will be fed to the reactors *via* two superhe

3.2.2 Mass velocities in reaction tubes

The number of tubes in the reactor depends on the mass rate of flow of alcohol or reaction products permitted. This in turn depends on the allowable pressure drop and the necessity to ensure turbulent flow conditions in the packed tubes. In order to keep the pressure drop to a minimum and obtain turbulent flow conditions the allowable mass rate of flow through the reactor tube will be based on the packed bed modified Reynolds number[1,2]

$$Re' = \frac{D_p G}{\mu(1-\varepsilon)} \geqslant 1000 \qquad (3.7)$$

where D_p is the diameter of the particle, 0.32 cm, and ε is the porosity of the packed bed. For random packed cylinders of the same size $\varepsilon = 0.393$.
Then for $Re = 1000$ the permissible mass rate of flow is

$$G = \frac{1000 \times 1.876 \times (1-0.393) \times 10^{-5}}{3.175 \times 10^{-3}} \text{ kg/m}^2\text{s}$$

$$= 3.588 \text{ kg/m}^2\text{s}$$

3.2.3 Conversion yield

The definition of the problem states that the conversion of the alcohol was to be 90% and it is desirable to consider the feasibility of this conversion in respect of the data available. Thus, it is well known that the dehydrogenation reaction is reversible and the equilibrium constant is a function of temperature; see equation (3.5). Furthermore the stoichiometric equation Figure 3.2 can be written for a conversion "x", thus:

$$A \rightleftharpoons K + H$$
$$1 - x \rightleftharpoons x + x$$

and the partial pressure of each component in the system will then be

$$p_A = \frac{\Pi(1-x)}{1+x} \quad : \quad p_K = \frac{\Pi x}{1+x} \quad : \quad p_H = \frac{\Pi x}{1+x}$$

and the equilibrium equation will be

$$K = \frac{\Pi x^2}{(1+x)(1-x)} = \frac{\Pi x^2}{1-x^2} \qquad (3.8)$$

where Π is the total pressure of the system

Combining equations (3.5) and (3.8) shows that the equilibrium conversion at different temperatures will be as shown in Table 3.2, which shows that if equilibrium was attained at the reactor exit the temperature of the reaction products would have to be between 300°C and 400°C. Since equilibrium conversion increases with temperature and in an actual reactor point-equilibrium will not be attained the reactor exit temperature will be estimated to be 400°C and this will form a basis for the design.

Table 3.2 – *Effect of temperature on equilibrium conversion.*

Temperature °C	Equilibrium Constant (Bar)	Equilibrium Conversion (Mole Fraction)
200	1.0127	0.58
300	14.4777	0.87
400	97.656	0.98
500	413.8091	0.99

3.2.4 *Treatment of reaction products*

The gaseous reaction products discharged from the reactor at a minimum of 400°C will be passed to the thermosyphon reboiler in order to initiate the preheating of the alcohol reactor feed. The reaction products will be cooled to near saturation in this reboiler so that they will enter the water cooled condenser at 125°C. There about 80% of the MEK and alcohol will be condensed and pumped to storage while the remainder will leave this unit as a saturated vapour in the noncondensable hydrogen. This vapour will be fed to the base of a packed adsorption column where the MEK and alcohol will be absorbed in water. The water will be recycled from the extraction column and its rate will be controlled to provide an aqueous effluent containing 10.0% MEK which will be fed to the extraction column. The hydrogen discharged from the top of the absorber will be dried and then fed into the furnace fuel system.

The aqueous effluent from the absorber will be pumped into an extraction column where it will be contacted with 1:1:2 trichlorethane to extract the MEK and alcohol. In excess of 95% of the MEK and alcohol will be extracted and the remainder will be recirculated back to the absorber for further treatment of the hydrogen/vapour mixture discharged from the condensers. The trichlorethane extract phase will be pumped to a distillation unit for separation of the solvent which will be recycled. The distillate from this column will be MEK and alcohol which will be mixed with the liquid product from the condenser and this will be treated in a MEK product still for the purification of the MEK. The alcohol discharged from the bottom of this column will be recycled back to the alcohol feed tank. The MEK product will be cooled and stored.

3.3 The material balance

The complete material balance over each unit will not be presented at this stage because the actual exit concentrations from a number of the equipment items depend on some form of step-wise calculation involving the actual design of the equipment. Consequently, these compositions will not be known at this stage and the complete material balance for the whole process will be assembled when they can be abstracted from the designs of the individual units. This has been done in the form of a summary and is presented in Figure 3.3.

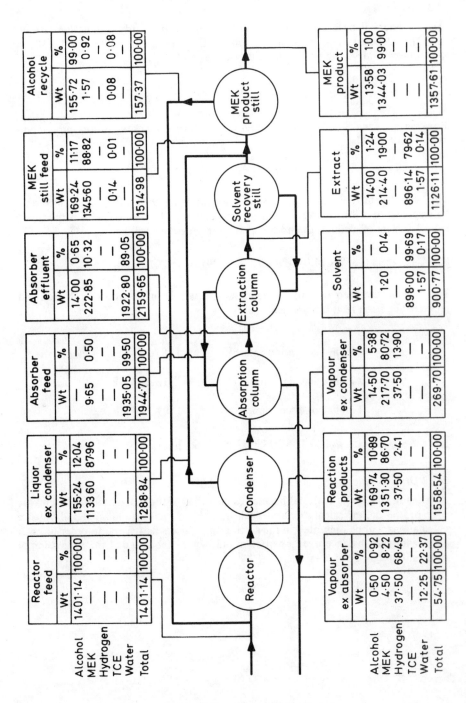

Figure 3.3 – *Material flowsheet.*

3.3.1 *Alcohol requirement*

The material balance over each unit will be based on the hourly production rate and 8000 h will be taken as the "on stream time". For a production rate of 1.0×10^7 kg per annum of MEK the production rate will be

$$(1.0 \times 10^7)/(8.0 \times 10^3) = 1.25 \times 10^3 \text{ kg/h}$$

However allowing 8.0% for spillage and other process losses the hourly production rate of MEK will be

$$(1.25 \times 10^3 \times 1.08) = 1.35 \times 10^3 \text{ kg/h}$$

This will be in the form of a solution containing 99% MEK, and since this corresponds to a 90% conversion in the reactor and 1.0 mole of MEK is produced from 1.0 mole of alcohol, the hourly alcohol requirement will be

$$(1350 \times 74.0)/(0.9 \times 72.0 \times 0.99) = 1557.24 \text{ kg/h}$$

This will be the feed rate to the reactors and will be made up of fresh feed and recycle. The fresh feed requirement will be

Alcohol converted to MEK	= 1387.5 kg/h
Alcohol leaving as product	= 13.64 kg/h
Total alcohol consumed	= 1401.14 kg/h

and the recycle rate will be 156.10 kg (alcohol) per hour.

3.3.2 *Other balances*

As already stated the material balance calculations of the individual plant items are part of the design calculations of each unit. Hence subsequent material balance calculations will be presented in the sections on process design and in order to assist the reader each balance over a particular unit will be tabulated at the conclusion of each unit design.

Chapter 4

PROCESS DESIGN – REACTOR

The general considerations for the design of the reactor were discussed in the previous Chapter and it was concluded there that a shell and tube type reactor was the most suitable for the duty envisaged. Furthermore it was decided that the heat of reaction would be supplied by flue gas and that, in order to achieve a 90% conversion, the exit reaction temperature must not fall below 400°C. On the basis of the flow rates, concentrations and conversions specified in Section 3.3.1 an overall material balance on the reactor would be that presented in Table 4.1 and this will be the basis for the design of the reactor.

Table 4.1 – *Reactor mass balance.*

	INPUT			OUTPUT	
Stream	Weight kg/h	%	Stream	Weight kg/h	%
MEK	1.30	0.08	MEK	1351.30	86.70
Hydrogen	–	–	Hydrogen	37.50	2.41
Alcohol	1557.24	99.92	Alcohol	169.74	10.89
Total	1558.54	100.00	Total	1558.54	100.00

4.1 Estimation of reaction rate

The catalytic dehydrogenation of 2-butanol to MEK has been widely studied and the findings of the reaction kinetic and chemical equilibrium researches have been summarised in Chapter 3 of this report. In all these investigations it has been found that mass transfer has a considerable effect on the rate of reaction, necessitating the use of the point-temperature and the partial pressures of each component at the catalyst surface in the reaction rate equation.

4.1.1 *Calculation of interface temperature*

The point temperature of the components at the interface depends upon the reaction rate and the rate of heat transfer from the bulk phase to the catalyst surface since the reaction is endothermic. This rate of heat transfer may be obtained from the definition of the heat transfer coefficient, thus,

$$Q = h_i A (T - T_i) \qquad (4.1)$$

The heat flux is determined from the heat absorbed during the reaction which takes place on the catalyst surface,

$$Q/A = \Delta H r_c^{(1)} \tag{4.2}$$

where the first approximation to the reaction rate, $r_c^{(1)}$ is obtained from equation (3.1) based on bulk phase temperature and component partial pressures. An estimation of the heat transfer coefficient was made employing the analogy between heat and momentum transfer. The j_H factor may be determined from a correlation with Reynolds number[13]

$$j_H = 1.06(Re')^{-0.41} \text{ for } Re' > 350 \tag{4.3}$$

The heat transfer factor is defined

$$j_H = (h_i/C_{pm}G)Pr_f^{2/3} \tag{4.4}$$

where Pr_f is the Prandtl Number, $(C_{pm}\mu_{mf}/k_m)$ evaluated at the film temperature $T_f = \frac{1}{2}(T_i + T)$. The interface temperature may then be determined by substitution of Q/A and h_i from equations (4.2) and (4.4) into equation (4.1), thus,

$$T_i = T - (\Delta H r_c^{(1)}/C_{pm}Gj_H)Pr_f^{2/3} \tag{4.5}$$

Evaluation of T_i requires an iterative calculation since it depends on the reaction rate and the procedure is summarised in Section 4.3.

4.1.2 *Evaluation of interface partial pressures*

The partial pressure of each component depends on the alcohol conversion, and was shown in Section 3.2.2 to be

(i) Partial pressure of 2-butanol $= \dfrac{\Pi(1-x)}{(1+x)} = p_A$

(ii) Partial pressure of MEK $= \dfrac{\Pi x}{(1+x)} = p_K$

(iii) Partial pressure of hydrogen $= \dfrac{\Pi x}{(1+x)} = p_H$

where Π is the total operating pressure of the reactor.

Furthermore the rate of mass transfer of each component depends on the partial pressure driving force and therefore on the bulk phase partial pressure and is related to the catalyst surface partial pressure through the rate of dehydrogenation and the rate of mass transfer. Hence evaluation of the gas/solid interface partial pressures for a designated conversion must involve the calculation of a preliminary rate of reaction based on the temperature at the catalyst surface employing bulk phase partial pressures. This rate must then be used to obtain a first approximation of the interface partial pressures from the rate of mass transfer since, at steady state, the point rate of reaction equals the point rate of mass transfer. Following this a new rate (reaction and mass transfer) can be estimated and this trial and error procedure repeated until the "rate" remains constant.

This procedure involves lengthy calculations and is conveniently performed by computer. Hence the following calculations are presented in a form that may be readily transformed into a program. Thus, the second approximation to the reaction rate is, by equation (3.1)

$$r_c^{(2)} = \frac{C_i[p_A - (p_K p_H/K_i)]}{p_K[1 + K_{Ai}p_A + K_{AKi}(p_A/p_K)]} \qquad (4.6)$$

where initially the pressures are bulk phase pressures to start the iterations. The rate $r_c^{(2)}$ is also the first estimate of the mass transfer rate which can be evaluated through the mass transfer factor j_D. This factor is related to the Reynolds number for beds packed with granulated packing and may be expressed $Re = d_p G/\mu_m$ where

d_p is the diameter of the catalyst packing
μ_m is the mean viscosity of the vapour
G is the mass velocity based upon the cross sectional area of the bed.

The j_D factors have been correlated with this Reynolds number [13] thus:

$$j_D = 0.99(Re')^{-0.41} \text{ for } Re' > 350 \qquad (4.7)$$

The procedure for evaluating the interfacial partial pressure of each component is identical and therefore the procedure will be described for the alcohol component. The mass transfer factor is defined

$$j_D = k_g M_m p_{fA} Sc_f^{2/3} G^{-1} \qquad (4.8)$$

where k_g is the mass transfer coefficient
M_m is the mean molecular weight
Sc_f is the Schmidt Number for the component evaluated at the mean film temperature $T_f = \tfrac{1}{2}(T_i + T)$.
p_{fA} is the film pressure factor expressed by

$$p_{fA} = \frac{(\Pi + \delta_A p_A) - (\Pi + \delta_A p_{Ai})}{\ln\left[(\Pi + \delta_A p_A)/(\Pi + \delta_A p_{Ai})\right]} \qquad (4.9)$$

from the stoichiometry of the reaction and

$$\delta_A = \left(\frac{\text{Number of moles, products} - \text{Number of moles, reactants}}{\text{Number of moles, component}}\right)$$

$$= \left(\frac{2-1}{1}\right) = 1 \qquad (4.10)$$

Rearranging equation (4.8) and substituting for p_{fA} from equation (4.9) gives

$$k_g \frac{(\Pi + p_A) - (\Pi + p_{Ai})}{\ln\left[(\Pi + p_A)/(\Pi + p_{Ai})\right]} = \frac{j_D G}{M_m} Sc_f^{-2/3}. \qquad (4.11)$$

Now the rate of mass transfer per unit area is

$$r_c^{(2)} = k_g(p_A - p_{Ai}) \qquad (4.12)$$

or

$$k_g = r_c^{(2)}/(p_A - p_{Ai}) \qquad (4.13)$$

so that k_g may be eliminated from equation (4.11), thus

$$\left(\frac{r_c^{(2)}}{p_A - p_{Ai}}\right) \frac{(\Pi + p_A) - (\Pi + p_{Ai})}{\ln\left[(\Pi + p_A)/(\Pi + p_{Ai})\right]} = \frac{j_D G}{M_m} Sc_f^{-2/3}. \qquad (4.14)$$

Simplifying equation (4.14) to express the interface partial pressure explicitly,

$$p_{Ai} = (\Pi + p_A) \exp(-c_A) - \Pi \qquad (4.15)$$

where

$$c_A = r_c^{(2)} M_m Sc_f^{2/3}/j_D G$$

Similar expressions for the interface partial pressures of the ketone and hydrogen products may be derived except that in these areas, the driving force is from the catalyst surface to the bulk phase and the direction of mass transfer is consequently reversed, thus,

$$p_{Ki} = \Pi - (\Pi - p_K) \exp(-c_K) \qquad (4.16)$$

$$p_{Hi} = \Pi - (\Pi - p_H) \exp(-c_H) \qquad (4.17)$$

The algorithm for calculation of the reaction rate based on the temperature and partial pressures of the components at the catalyst surface may now be summarised as follows,

(i) Calculate $r_c^{(1)}$ from equation (3.1)
(ii) Calculate T_i from equation (4.5)
(iii) Calculate $r_c^{(2)}$ from equation (4.6)
(iv) Evaluate p_{Ai}, p_{Ki} and p_{Hi} from equations (4.15), (4.16) and (4.17) respectively using $r_c^{(2)}$
(v) Recalculate $r_c^{(2)}$ from equation (4.6) and iterate using successive values of interface partial pressures [from equations (4.15), (4.16) and (4.17)] and interface temperature [from equation (4.5)] until $r_c^{(2)}$ converges.

The rate of reaction must be calculated at each increment along the reactor tube in order to ascertain the reactor length and the above algorithm was incorporated as a sub-routine in the main reactor design model described below. However, before making a rigorous design of the reactor the approximate dimensions will be established.

4.2 Approximate dimensions of reactor

The rigorous calculation of even one increment of reactor tube length is impractical manually because of the complexity of the analysis and the large number of iterations involved. However it is desirable to obtain an order of magnitude estimate of the reactor dimensions before the rigorous design is undertaken. This will check the rigorous model and give a preliminary estimate of magnitude of the variables, and indicate the number and size of the increments required in the iterative calculations.

4.2.1 *Approximate estimate of reactor tube length*

Perona & Thodos[3] made use of the "Height of a Reactor Unit" concept to assess their reactor and their correlation for the HRU with temperature and flow rate in SI units would be

$$HRU \text{ (metres)} = 224.9G/(140.89T - 86168)^{5/6} \qquad (4.18)$$

where G is the flow rate in kg/m²s and, by Section 3.2.2, $G = 3.588$ kg/m²s and T is the catalyst surface temperature in kelvins. Initially, let the mean catalyst surface temperature in the reactor be 400°C. Then,

$$HRU = 224.9 \times 3.588/[(140.89 \times 673) - 86168]^{5/6} = 0.422 \text{ metres}$$

The reaction equilibrium constant at the estimated catalyst surface temperature (400°C) is, by equation (3.5)

$$\log K = (2790/673) + 1.510 \log 673 + 1.871$$

or

$$K = 99.0 \text{ (see Table 3.2)}$$

Then the composition of the reaction products will be, by equation (3.1) at this equilibrium ($r_c = 0$)

$$\frac{C[p_A - p_K p_H/K]}{p_K[1 + K_A p_A + K_{AK}(p_A/p_K)]} = 0$$

or

$$p_A - (p_K p_H/K) = 0 \qquad (4.19)$$

Substituting from Section 3.2.3

$$p_K = p_H = \Pi x/(1+x) \text{ and } p_A = \Pi(1-x)/(1+x)$$

gives

$$99.0 \left(\frac{1-x}{1+x}\right) - \frac{\Pi x^2}{(1+x)^2} = 0$$

For an operating pressure of 2.0 bar

Then
$$99.0(1-x)^2 - 2.0x^2 = 0 \quad \text{or} \quad x = 0.990$$

and
$$p_A^* = 2.0(1-0.990)/(1+0.990) = 0.010$$

$$p_K = p_H = (2.0 \times 0.990)/(1+0.990) = 0.9949$$

so that the required length of reactor Z will be

$$Z = (HRU) \ln[(p_{A1} - p_A^*)/(p_{A2} - p_A^*)] \tag{4.20}$$

$$= 0.422 \ln[(2.0-0.010)/(p_{A2}-0.010)] \tag{4.21}$$

Equation (4.21) was solved to obtain the relation between Z and p_A and therefore Z and the conversion x. The results obtained are plotted in Figure 4.1 and presented in Table 4.2 where it can be seen that for a fractional conversion of 90% alcohol the height of catalyst in the reactor tubes should be 1.3 metres.

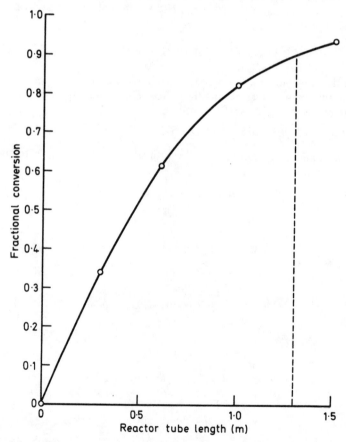

Figure 4.1 — *Variation of fractional conversion with reactor tube length (HRU approach).*

Table 4.2 — *Relation between conversion and height of catalyst in reactor tube.*

Reactor Tube Length Z(m)	Alcohol Partial Pressure p_A (bar)	Conversion $x = (\pi - p_A)/(\pi + p_A)$
0.0	2.000	0.0
0.30	0.987	0.339
0.61	0.478	0.614
1.00	0.196	0.821
1.52	0.064	0.937
2.13	0.0227	0.997

Approximate length of reactor tubes: 1.3 m

4.3 Estimation of reactor tube diameter

Jenson & Jeffreys[8] have described a method of estimating the diameter of a catalyst reactor tube from thermal considerations. Their model leads to the partial differential equation (3.6). For the purpose of a preliminary estimate of the reactor tube diameter, the temperature drop between the entrance and exit of the reactor will be assumed to be linear. The equation (3.6) becomes

$$\frac{d^2 T}{dx^2} + \frac{1}{x} \frac{dT}{dx} - \left[\frac{GC_p}{k_E} \left(\frac{\Delta T}{\Delta z} \right) + \frac{v \Delta H r_c}{k_E} \right] = 0 \qquad (4.22)$$

Let

$$\left[\frac{GC_p}{k_E} \left(\frac{\Delta T}{\Delta Z} \right) + \frac{v \Delta H r_c}{k_E} \right] = P$$

Then along the tube axis, at $x = 0$; $dT/dx = 0$ from symmetry considerations. Now taking the Laplace Transformation of equation (4.22) gives

$$\frac{d\bar{T}}{ds} + \frac{1}{s} \bar{T} = -\frac{P}{s^4} \qquad (4.23)$$

The integrating factor for equation (4.23) is

$$IF = \exp \int \frac{ds}{s} = s.$$

or

$$\frac{d}{ds}(s\bar{T}) = -\frac{P}{s^3} \qquad (4.24)$$

Integration of equation (4.24) gives

$$\bar{T} = P/2s^3 + C/S$$

which on inverting leads to

$$T = \tfrac{1}{4}Px^2 + C \tag{4.25}$$

Consideration of Perona & Thodos' paper shows that the reaction rate is very slow below 300°C and this will be set as the lowest temperature at the tube axis. That is at $x = 0$: $T = 573$ K

The reactor is to be designed for a conversion of 90%. Thus for this calculation the reactor tube diameter will be estimated 45% conversion i.e. $x = 0.45$.

Then

$$p_A = \Pi(1-x)/(1+x) = 2.0(1-0.45)/(1+0.45) = 0.758 \text{ bar}$$

$$p_K = p_H = \Pi x/(1+x) = 2.0 \times 0.45/(1+0.45) = 0.62 \text{ bar}.$$

Let $T = 450°C$ (= 723 K), then

$$\log K = (-2790/723) + 1.510 \log 723 + 1.865 \quad \text{or} \quad K = 210.5 \text{ bar}$$

$$\log C = (-5964/723) + 8.464 \quad \text{or} \quad C = 1.641 \text{ kg mole}/m^2 h$$

$$\log K_A = (-3422/723) + 5.326 \quad \text{or} \quad K_A = 3.916 \text{ bar}$$

$$\log K_{AK} = (269.2/723) - 0.1959 \quad \text{or} \quad K_{AK} = 1.501$$

Then at $x = 0.45$

$$r_c = \frac{1.641[0.758 - (0.620)^2/210.5]}{0.62[1.0 + (3.916 \times 0.758) + (1.501 \times 0.758/0.620)]}$$

$$= 0.3448 \text{ kg mole/h m}^2$$

$$= 0.957 \times 10^{-4} \text{ kg mole}/m^2 \text{ s}$$

Although the calculated rate of reaction which is based on bulk phase conditions of temperature and partial pressures provide an initial estimate, the actual rate will be much lower and a value of $r_c = 0.25 \times 10^{-4}$ kg mole/m^2s will be assumed for evaluation of the minimum tube radius.

The heat of reaction is a function of temperature and has been correlated with temperature by the method suggested in Cooper & Jeffreys[14]. The method and the correlation is presented in Appendix B and the heat of reaction is

$$H = 73900 \text{ kJ per kg mole at 723 K}$$

The effective thermal conductivity k_E has been estimated in Appendix C to be 0.385 W/m K. The surface area of the catalyst is $6/d_p$ which for right cylindrical particles 0.32 cm diameter gives $s = (6.0 \times 10^3)/0.32 = 1889$ m^2/m^3 of particles. For a random packed bed of cylindrical particles the porosity is $\varepsilon = 0.393$ Then the surface area per unit volume of packed bed will be:

$$v = s(1-\varepsilon) = 1889(1-0.393)$$
$$= 1147 \text{ m}^2/\text{m}^3 \text{ of bed}$$

The temperature drop along the length of the reactor has been estimated to be 100 K and therefore

$$(\Delta T/\Delta Z) = -100/2.0 = -50 \text{ K/m}.$$

The heat capacity of the reaction mixture at 45% conversion has been estimated from the heat capacity of each component employing the expressions presented in Appendix A. At 723 K the heat capacity of MEK is 44.21 cal/g-mole K: the heat capacity of 2-butanol is 49.71 cal/g-mole K. the heat capacity of hydrogen is 7.05 cal/g-mole K and the weighted mean heat capacity of the mixture is

$$1.603 \text{ cal/gK} (= 6.706 \text{ kJ/kgK})$$

Then

$$P = \frac{3.588 \times 6.706 \times 10^3 \times (-50)}{0.385} + \frac{1147 \times 73900 \times 0.25 \times 10^{-4}}{0.385 \times 10^{-3}}$$

$$= 1.955 \times 10^6$$

Substitution in equation (4.25) gives

$$773 = \tfrac{1}{4}(1.955 \times 10^6)a^2 + 573 \tag{4.26}$$

where a is the radius of the reaction tube. From equation (4.26) the tube radius is:—

$$a = (4.091 \times 10^{-4})^{0.5} \text{ m} = 2.02 \times 10^{-2} \text{ m} = 2.02 \text{ cm, for safety say 2.1 cm}$$

That is Reactor Tube Diameter = 4.2 cm

This value of the reactor tube diameter will be applied in the subsequent design calculations.

4.4 Mathematical model for conversion and temperature profiles in reactor tubes

The model employed here is a modified version of that developed by Jenson & Jeffreys (8). Consider an element of volume of reactor tube (Figure 4.2). The 2-butanol reacting in volume element = $2\pi r \delta r \delta z v r_c$ where v is the surface area per unit volume of catalyst. Heat absorbed by endothermic reaction in element = $2\pi r \delta r \delta z v r_c \Delta H$. Consider a mass balance taking into account bulk flow, radial diffusion and reaction.

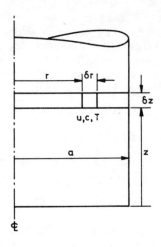

Figure 4.2 — *Element of volume of reactor tube.*

$$2\pi r \delta r u c - 2\pi r \delta z \frac{D_E \partial(uc)}{u \, \partial r} - \left[2\pi r \delta r u c + \frac{\partial}{\partial z}(2\pi r \delta r u c) \delta z \right]$$
$$+ \left[2\pi r \delta r u c - 2\pi r \delta z \frac{D_E \partial(uc)}{u \, \partial r} + \frac{\partial}{\partial r}\left(2\pi r \delta z \frac{D_E \partial(uc)}{u \, \partial r} \right) \delta r \right] = 2\pi r \delta r \delta z v r_c$$

where D_E/u is the ratio of effective diffusivity to linear velocity in the tube. Hence

$$\frac{\partial(uc)}{\partial z} - \frac{D_E}{u}\left(\frac{\partial^2(uc)}{\partial r^2} + \frac{1}{r}\frac{\partial(uc)}{\partial r} \right) + v r_c = 0 \qquad (4.27)$$

now fractional conversion,

$$x = (u_0 c_0 - uc)/u_0 c_0 \qquad (4.28)$$

$$\therefore \partial(uc) = -u_0 c_0 \partial x \qquad (4.29)$$

Substitution of $\partial(uc)$ from (4.29) into equation (4.27) gives

$$\frac{\partial x}{\partial z} - \frac{D_E}{u}\left(\frac{\partial^2 x}{\partial r^2} + \frac{1}{r}\frac{\partial x}{\partial r} \right) - \frac{v r_c}{u_0 c_0} = 0 \qquad (4.30)$$

Similarly, a heat balance over the volume element yields:

$$\frac{\partial T}{\partial z} - \frac{k_E}{G C_p}\left(\frac{\partial^2 T}{\partial r^2} + \frac{1}{r}\frac{\partial T}{\partial r} \right) + \frac{\Delta H v r_c}{G C_p} = 0 \qquad (4.31)$$

In order to ease manipulation of equations (4.30) and (4.31), the following constants are introduced.

Let $k_E/GC_p = \alpha$; $\Delta Hv/GC_p = \beta$

$D_E/u = \gamma$; $v/u_0c_0 = \phi$

Equations (4.30) and (4.31) then become

$$\frac{\partial T}{\partial z} - \alpha\left(\frac{\partial^2 T}{\partial r^2} + \frac{1}{r}\frac{\partial T}{\partial r}\right) + \beta r_c = 0 \quad (4.32)$$

$$\frac{\partial x}{\partial z} - \gamma\left(\frac{\partial^2 x}{\partial r^2} + \frac{1}{r}\frac{x}{r}\right) - \phi r_c = 0 \quad (4.33)$$

Consider the *Boundary Conditions* for the partial differential equations.

For Mass Transfer [equation (4.33)]

(i) $z = 0$; $x_0 = 0.00085$ (due to recycle ketone) $\quad (4.34)$

(ii) $r = 0$; $\frac{\partial x}{\partial r} = 0$ (by symmetry) $\quad (4.35)$

(iii) $r = a$; $\frac{\partial x}{\partial r} = 0$ (no mass transfer through tube wall) $\quad (4.36)$

For Heat Transfer [equation (4.32)]

(i) $z = 0$; $T_0 = 773$ K (feed inlet temperature) $\quad (4.37)$

(ii) $r = 0$; $\frac{\partial T}{\partial r} = 0$ (by symmetry) $\quad (4.38)$

(iii) This boundary condition is determined by considering a heat balance over tube δz,

Heat flow in through tube wall = Heat flow out by conduction. Therefore,

$$U_D \pi a \delta z (T_{fg} - T_a) = k_E \left.\frac{\partial T}{\partial r}\right|_a \pi a \delta z$$

$$\left.\frac{\partial T}{\partial r}\right|_a = \frac{U_D}{k_E}(T_{fg} - T_a) \quad (4.39)$$

where T_{fg} is the point-temperature of the heating medium.

4.5 Numerical solution of partial differential equations

Initially, the explicit method of solution was attempted but was found to be mathematically unstable due to the small radial increment that was required to produce temperature and conversion profiles. Therefore the more accurate, implicit method of Crank and Nicholson was selected to overcome this difficulty by employing second order finite differences. Some stages of algebraic manipulation have been omitted in this section for the sake of brevity.

Expressing equation (4.32) in finite difference form:

$$\frac{T_{n,m+1}-T_{n,m}}{k} - \frac{\alpha}{2}\left[\frac{T_{n+1,m}-2T_{n,m}+T_{n-1,m}}{h^2} + \frac{T_{n+1,m}-T_{n-1,m}}{2nh^2}\right.$$
$$\left.+\frac{T_{n+1,m+1}-2T_{n,m+1}+T_{n-1,m+1}}{h^2} + \frac{T_{n+1,m+1}-T_{n-1,m+1}}{2nh^2}\right]+\beta r_c = 0$$

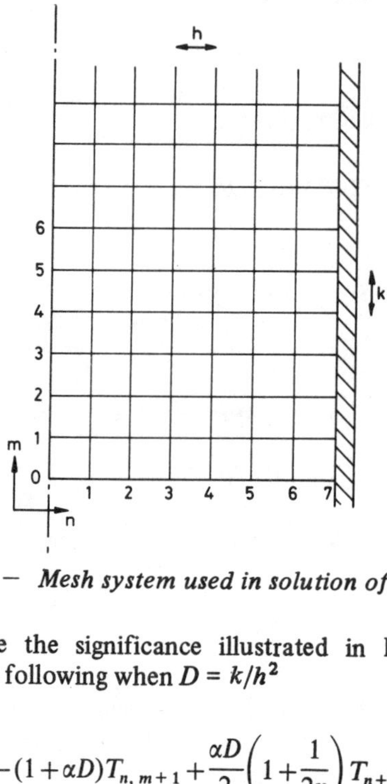

Figure 4.3 — Mesh system used in solution of equations.

where the symbols have the significance illustrated in Figure 4.3. The latter equation simplifies to the following when $D = k/h^2$

$$\frac{\alpha D}{2}\left(1-\frac{1}{2n}\right)T_{n-1,m+1}-(1+\alpha D)T_{n,m+1}+\frac{\alpha D}{2}\left(1+\frac{1}{2n}\right)T_{n+1,m+1}$$
$$= \beta kr_c - \frac{\alpha D}{2}\left(1-\frac{1}{2n}\right)T_{n-1,m}-(1-\alpha D)T_{n,m}-\frac{\alpha D}{2}\left(1+\frac{1}{2n}\right)T_{n+1,m} \quad (4.40)$$

Similarly from equation (4.33)

$$\frac{\gamma D}{2}\left(1-\frac{1}{2n}\right)x_{n-1,m+1}-(1+\gamma D)x_{n,m+1}+\frac{\gamma D}{2}\left(1+\frac{1}{2n}\right)x_{n+1,m+1}$$
$$= -\phi kr_c - \frac{\gamma D}{2}\left(1-\frac{1}{2n}\right)x_{n-1,m}-(1-\gamma D)x_{n,m}-\frac{\gamma D}{2}\left(1+\frac{1}{2n}\right)x_{n+1,m} \quad (4.41)$$

However, these equations cannot be employed for conditions at the tube axis since an indeterminate fraction arises because from equation (4.38) $\partial T/\partial r = 0$ when $r = 0$ therefore $(1/r)(\partial T/\partial r)$ is indeterminate. This difficulty may be surmounted by the use of L'Hôpital's Rule which states that

$$\lim_{r \to 0} \frac{1}{r} \frac{\partial T}{\partial r} = \frac{\partial^2 T}{\partial r^2}$$

Equation (4.32) then becomes

$$\frac{\partial T}{\partial z} - 2\alpha \frac{\partial^2 T}{\partial r^2} + \beta r_c = 0 \qquad (4.42)$$

Expressing equation (4.42) in finite difference form at $n = 0$, and from equation (4.38) which implies that $T_{-1,m} = T_{1,m}$ gives

$$(1+2\alpha D)T_{0,m+1} - 2\alpha D T_{1,m+1} = -\beta k r_c + (1-2\alpha D)T_{0,m} + 2\alpha D T_{1,m} \qquad (4.43)$$

Similarly from equation (4.33)

$$(1+2\gamma D)x_{0,m+1} - 2\gamma D x_{1,m+1} = \phi k r_c + (1-2\gamma D)x_{0,m} + 2\gamma D x_{1,m} \qquad (4.44)$$

In order to solve the equations for the axis points, it is necessary to eliminate $T_{0,m+1}$ which is completed by writing equation (4.40) for $n = 1$ and substituting for $T_{0,m+1}$ from equation (4.43) to give:

$$\left[\frac{\alpha^2 D^2}{2(1+2\alpha D)} - (1+\alpha D)\right] T_{1,m+1} + \frac{3\alpha D}{4} T_{2,m+1} = \left[1 + \frac{\alpha D}{4(1+2\alpha D)}\right] \beta k r_c$$

$$-\frac{\alpha D}{4}\left[1 + \frac{(1-2\alpha D)}{(1+2\alpha D)}\right] T_{0,m} - \left[(1-\alpha D) + \frac{\alpha^2 D^2}{2(1+2\alpha D)}\right] T_{1,m} - \frac{3\alpha D}{4} T_{2,m} \qquad (4.45)$$

The same procedure was followed for the mass transfer equation yielding:

$$\left[\frac{\gamma^2 D^2}{2(1+2\gamma D)} - (1+\gamma D)\right] x_{1,m+1} + \frac{3\gamma D}{4} x_{2,m+1} = -\left[1 + \frac{\gamma D}{4(1+2\gamma D)}\right]\phi k r_c$$

$$-\frac{\gamma D}{4}\left[1 + \frac{(1-2\gamma D)}{(1+2\gamma D)}\right] x_{0,m} - \left[(1-\gamma D) + \frac{\gamma^2 D^2}{2(1+2\gamma D)}\right] x_{1,m} - \frac{3\gamma D}{4} x_{2,m} \qquad (4.46)$$

The equations for use on the tube wall necessitated the introduction of a fictitious temperature on the outside of the tube and thus the general equations, (4.40) and (4.41), were used in conjunction with the boundary conditions at the tube wall to eliminate the fictitious points. Thus, expressing the boundary condition described by equation (4.39) in finite difference form

$$\frac{T_{8,m} - T_{7,m}}{h} = \frac{U_D}{k_E}(T_{fg} - T_{7,m}) \qquad (4.47)$$

also for the $(m+1)$ increment

$$\frac{T_{8,m+1} - T_{7,m+1}}{h} = \frac{U_D}{k_E}(T_{fg} - T_{7,m+1}) \tag{4.48}$$

combining equations (4.48) and (4.49) to give second order differences at a point midway between the m and $(m+1)$ increments.

$$T_{8,m+1} + T_{8,m} = \frac{2hU_D}{k_E}T_{fg} + \left(1 - \frac{hU_D}{k_E}\right)(T_{7,m+1} + T_{7,m}) \tag{4.49}$$

The fictitious temperature points $T_{8,m+1}$ and $T_{8,m}$ are eliminated between equation (4.40) written for $n = 7$ and equation (4.49) to yield the temperature at the wall, $T_{7,m+1}$, thus

$$\left[\frac{\alpha D}{2}\left(1 + \frac{1}{14}\right)\left(1 - \frac{hU_D}{k_E}\right) - (1+\alpha D)\right]T_{7,m+1} = \beta k r_c - \frac{hU_D}{k_E}\alpha D T_{fg}$$

$$-\left[\frac{\alpha D}{2}\left(1 + \frac{1}{14}\right)\left(1 - \frac{hU_D}{k_E}\right) + (1-\alpha D)\right]T_{7,m}$$

$$-\left[\frac{\alpha D}{2}\left(1 - \frac{1}{14}\right)\right](T_{6,m+1} + T_{6,m}) \tag{4.50}$$

Then eliminating $T_{7,m+1}$ by substitution from equation (4.50) into equation (4.40) which has been written for $n = 6$ produces an equation of the following form;

$$a_1 T_{5,m+1} + a_2 T_{6,m+1} = a_3 \beta k r_c + a_4 T_{fg} + a_5 T_{5,m} + a_6 T_{6,m} + a_7 T_{7,m}$$

The coefficients of terms in the above equation are complex and are presented in expanded form in Table 4.3.

Table 4.3 – *Co-efficients of temperature equations for use at the tube wall.*

Coefficient	"Expanded form"
a_1	$-\{[0.2129 + 0.2455\,(hU_D/k_E)]\alpha^2 D^2 + 0.4583\,\alpha D\}$
a_2	$\{[0.2129 + 0.5357\,(hU_D/k_E)]\alpha^2 D^2 + [1.4643 + 0.5367\,(hU_D/k_E)]\alpha D + 1\}$
a_3	$-\{[1.0059 + 0.5357\,(hU_D/k_E)]\alpha D + 1\}$
a_4	$0.5802\,(hU_D/k_E)\alpha^2 D^2$
a_5	$\{[0.2129 + 0.2455\,(hU_D/k_E)]\alpha^2 D^2 + 0.4583\,\alpha D\}$
a_6	$-\{[0.2129 + 0.5357\,(hU_D/k_E)]\alpha^2 D^2 + 0.5357\,(1 - hU_D/k_E)\alpha D - 1\}$
a_7	$1.0832\,\alpha D$

The equivalent manipulation for the mass transfer equation is somewhat simpler. It involved writing equation (4.41) for $n = 7$ and using equation (4.35) to produce $x_{6,m} = x_{8,m}$ and $x_{6,m+1} = x_{8,m+1}$ thus yielding by substitution

$$\gamma D x_{6,m+1} - (1+\gamma D)x_{7,m+1} = -\phi k r_c - \gamma D x_{6,m} - (1-\gamma D)x_{7,m} \quad (4.51)$$

In order to test if the desired fractional conversion of 2-butanol (90%) had been achieved, it was necessary to evaluate the mean conversion from the conversion profile, and, in addition, the mean temperature and conversion was required for calculation of the reaction rate. Hence the mean conversion is given by:

$$\bar{x} = \int_0^a rx\, dr / \int_0^a r\, dr \quad (4.52)$$

The numerator of equation (4.52) may be deduced using Simpson's rule

$$\int_0^a rx\, dr = \frac{1}{36} a^2 (x_1 + 4x_2 + 2x_3 + 4x_5 + 2x_6 + 4x_7 + x_8)$$

The denominator of equation (4.52) is given by

$$\int_0^a r\, dr = \tfrac{1}{2} a^2,$$

Therefore

$$\bar{x} = \frac{1}{18} (x_1 + 4x_2 + 2x_3 + 4x_5 + 2x_6 + 4x_7 + x_8) \quad (4.53)$$

Similarly for mean temperature:

$$\bar{T} = \frac{1}{18} (T_1 + 4T_2 + 2T_3 + 4T_5 + 2T_6 + 4T_7 + T_8) \quad (4.54)$$

Equations (4.53) and (4.54) were solved using a function segment due to their basic similarity.

The extremely large number of calculations inherent in the solution of the equations, due to the comprehensive procedure adopted for determination of the reaction rate at each length increment of the reactor tube, necessitated the use of a computer program. In addition to providing the solutions very rapidly, it was possible to evaluate the physical properties of the vapour components at the conditions of temperature and pressure prevailing at each point of the reactor tube length. This had obvious advantages since the mean temperatures of the feed and products differed by over 100 K resulting in significant changes in vapour properties. This is confirmed by the output of the program (Tables 4.9, 4.10 and 4.11) which list the properties of the system at fractional conversions of approximately 0.1, 0.5 and 0.9.

Although a detailed flow diagram is included in Appendix D which illustrates the logic sequences employed in the program, a general summary together with the more important features will be presented here. The program consisted of a master segment whose function was to input data, evaluate the coefficients of the temperature and conversion equations and then to output the radial profiles of temperature and fractional conversion at chosen length increments of 5 cm. The maximum longitudinal step length, k, was determined by writing the partial differential equations as first order differences as described by Jenson & Jeffreys[8]. The modulus of these equations was then calculated, so that no negative coefficients would arise, to ensure a stable calculation process,

For all positive coefficients, $(1 - 4\alpha D) > 0$ therefore $\alpha D > \frac{1}{4}$ now $\alpha = k/GC_{pm}$ = $(0.385 \times 10^{-3})/(3.588 \times 6.706) = 1.60 \times 10^{-5}$, therefore

$$D = \frac{k}{h^2} \frac{0.25}{1.60 \times 10^{-5}}$$

For a radial step length of 3 mm, $h = 3 \times 10^{-3}$ m, and

$$k < \frac{0.25 \times 9 \times 10^{-6}}{1.60 \times 10^{-5}} \text{ m that is } k < 0.140 \text{ m}$$

A longitudinal step length of 10mm was therefore considered to be suitable. Since the temperature of the flue gas changes as the reaction proceeds along the length of the catalyst tubes, this temperature was also re-calculated by a heat balance at length increments equal to the spacing of the baffles. The justification for this length is based on an assumption that the flue gas in each shell side compartment is well mixed. The details of the heat balance are given in Section 4.7 and the equation used to evaluate T_{fg} is as follows,

$$T_{fg(b+1)} = T_{fg(b)} + [(1557.2/74)(\bar{x}_{b+1} - \bar{x}_b)\Delta H - 1558.5 C_{pm}(\bar{T}_b - \bar{T}_{b+1})] RC_p' \quad (4.55)$$

The reaction rate is also sensitive to the total pressure in the reactor tube since all the components are present in the vapour phase. Therefore the pressure drop across each length increment was calculated using the equation described in Section 4.7.5,

$$\Delta P = 2fkG^2 v/\rho_m \, e^{1.7} \quad (4.56)$$

The new total pressure was then evaluated by successively subtracting the pressure drop for each increment.

Solution of the simultaneous set of equations for both temperature and fractional conversion was facilitated using elimination by the Tridiagonal method. The algorithm for this technique which was incorporated into the program as a subroutine is presented in Appendix E.

Calculations of the component physical properties using the equations developed in Appendices A,B,C,F and the reaction rate based upon interface partial pressures and temperature were also completed in a subroutine. The detailed procedure for evaluation of the rate of reaction has already been summarised in Section 4.1.2.

Finally, to take account of the variation in catalyst activity during the on-line period, a factor which is applied to the rate of reaction, based upon economic considerations, is derived in the following section.

4.6 Estimation of the catalyst activity factor

During the period of operation of the reactor the activity of the catalyst decreases and if the catalyst bed is designed on the basis of the initial activity the desired fractional conversion will not be achieved. However, if the length of the tubes are designed to accommodate the lowest value of catalyst activity, the conversion of 2-butanol will be in excess of 90% resulting in over-design of the separation equipment. Therefore, there is an optimal size of the bed which may be estimated by economic considerations, bearing in mind that the activity of the catalyst is not known to any degree of accuracy for the prevailing conditions in the reactor. Rudd & Watson[15] have presented a method for estimation of the overdesign factor which will be employed. Now let

Q' = required production (kg per day)
A_L = lowest possible limit of catalyst activity (kg per day per kg catalyst)
A_H = highest possible limit of catalyst activity
C_c = cost of catalyst (£ per kg)
C_s = cost of shutdown to rebuild reactor
D' = optimal size of catalyst bed.

then if success is achieved on the first design

$$C(D'A) = C_c D' \quad \text{if } D'A > Q'$$

but if redesign is required corresponding to a plant shutdown

$$C(D'A) = C_c D' + C_c[(Q/A) - D'] + C_s \quad \text{if } D'A < Q'$$

where $C_c [(Q/A) - D']$ is the extra cost for the correct amount of catalyst. The expected cost is then

$$\bar{C}(D') = \int_{A_1}^{Q'/D'} \left(\frac{C_c Q}{A} + C_s\right)\frac{dA}{(A_H - A_L)} + \int_{Q'/D'}^{A_H} (C_c D')\frac{dA}{(A_H - A_L)}$$

$$= \frac{1}{A_H - A_L}\left[C_c Q' \ln \frac{Q'}{D' A_L} + C_s\left(\frac{Q'}{D} - A_L\right) + C_c D'\left(A_H - \frac{Q'}{D'}\right)\right] \quad (4.57)$$

The minimum cost may then be found by differentiating equation (4.57) with respect to D' and setting the derivative equal to zero. Therefore the minimum expected cost is attained when

$$D^* = \tfrac{1}{2}\left\{\frac{Q'}{A_H} + \left[\left(\frac{Q'}{A_H}\right)^2 + \frac{4C_s Q'}{C_c A_H}\right]^{1/2}\right\} \quad (4.58)$$

where D^* is the recommended catalyst quantity based on the minimum expected cost. Taking the upper catalyst activity as an arbitrary base, the overdesign factor, f_o can be defined as

$$f_o = \frac{D^*}{(Q/A_H)}$$

Substituting this into equation (4.58) yields the optimum overdesign factor,

$$f_o' = \tfrac{1}{2}\left[1 + \left(1 + \frac{4C_s A_H}{C_c Q'}\right)^{1/2}\right] \quad \text{for } f_o' < 1/(1-\Delta) \qquad (4.59)$$

where Δ, the fractional uncertainty in the catalyst activity is given by:

$$\Delta = (A_H - A_L)/A_H. \quad \text{If } f_o' \geqslant 1/(1-\Delta) \text{ then:}$$

$$f_o' = \frac{1}{1-\Delta} \qquad (4.60)$$

now Ford & Perlmutter[5] reported that over a 15 h run under laboratory conditions the catalyst activity decreased to 77% of the initial activity (*i.e.* after regeneration by air oxidation followed by reduction with hydrogen). For much longer reaction times, only a small further reduction in catalyst activity occured. Therefore A_H = 1.0 as an arbitrary value and A_L = 0.77. Hence from equation (4.60) which is the optimum factor corresponding to the worst conditions for low uncertainty in the catalyst activity, where $\Delta = (1.0 - 0.77)/1.0 = 0.23$.

$$f_o' = 1/(1-\Delta) = 1/(1-0.23) = 1.3$$

It is now necessary to investigate the economic implications of a reactor modification using equation (4.59) to decide which overdesign factor to use.

The present price of (65/35) brass[16] = £822.5 per metric ton. As brass cylinders are likely to be a "one-off" commodity, it will be assumed that the actual cost of catalyst be four times the cost of the raw material.

Hence C_c = (822.5 × 4) 1000 £/kg = 3.29 £/kg.

If the reactors were not able to attain the required fractional conversion it would be necessary to replace them. Assuming that the reactors would be fabricated off-site, it was estimated that the maintenance schedule be of 10 days duration, then loss of MEK product = 10 × 24 × 1350 kg = 324000 kg. Price of MEK[17] is £254.5 per metric ton. Therefore Financial loss = £ (324000 × 254.5)/1000 = £82500. Therefore, the total cost of the shutdown will be assumed to be £100000 to take account of overheads, labour *etc*. So

$$C_s = £100\,000, \text{ also } Q' = 32\,400 \text{ kg MEK per day.}$$

Calculation of A_H. If a catalyst activity factor of unity is assumed, conversion of 90% is achieved in 2.0 m, so volume of reactor tube

$$= \tfrac{1}{4}\pi D^2 L \text{ m}^3 = \tfrac{1}{4}\pi(42 \times 10^{-3})^2 \times 2 \text{ m}^3 = 2.77 \times 10^{-3} \text{ m}^3.$$

If three reactors are required, each containing 100 tubes approximately:

Total volume of reactor tubes	=	0.831 m³
Voidage fraction of catalyst	=	0.393
So volume of catalyst	=	0.607 × 0.831 m³ = 0.504 m³
Density of (65/35) brass	=	8477 kg/m³
So mass of catalyst required	=	4272 kg
MEK production rate	=	1350 kg/h = 32400 kg/day

Therefore A_H = 7.584 kg MEK per day per kg catalyst.

From equation (4.59)

$$f'_0 = \tfrac{1}{2}\left[1+\left(1+\frac{4\times 100\,000 \times 7.584}{3.29 \times 32400}\right)^{1/2}\right]$$

$$= \tfrac{1}{2}[1+(1+28.5)^{1/2}]$$

$$= 3.2 > 1/(1-\Delta)$$

Comparison between the overdesign factors obtained using equations (4.59) and (4.60) clearly shows that the economic factors are not important if the catalyst activity is maintained between the specified limits. This implies that overdesign of the reactor tube length is essential since replacement of the reactors is not economically feasible. In fact the catalyst activity would have to be reduced to below 31% of the initial value before reactor replacement could be considered as a practical alternative. Therefore the overdesign factor based on the uncertainty of catalyst activity will be used in the computer program. Since activity changes affect the reaction rate, the overdesign factor was incorporated into the program as follows; Rate of Reaction = r_c/f'_0.

4.7 Estimation of reactor length based on heat transfer considerations.

4.7.1 Tube layout and shell dimensions

A preliminary estimate of reactor length employing the HRU approach (Section 4.2) showed that a 1.3 m reactor is required. However the estimate is considerably increased when mass transfer effects, and heat transfer in the tubes and reduction in catalyst activity is considered. From the output of the computer program, the length of the reactor tubes is found to be 3.0 m which will be assumed initially to be adequate for the required heat transfer. One tube pass was selected in order to minimise the pressure drop on the product stream.

Using 1½ inch nominal size tubes; (d_i = 40.9 mm, d_o = 48.3 mm) on a 60.3 mm triangular pitch; Flow area per tube = $\tfrac{1}{4}\pi d_i^2$ = $\tfrac{1}{4}\pi \times 0.0409^2$ m² = 1.3138 × 10^{-3} m². Total flow area required = M/G = 1558.54/(3588 × 3600) m² = 0.1206 m²

So number of tubes required = 92.

Using the design procedure described by Holland et al[18]; Maximum number of tubes on inside diameter of shell is, $n_d^2 = (4n_t-1)/3$. Initial estimate of total number of tubes will be assumed to be the sum of the number required for heat transfer and the number required for the insertion of six tie rods.

So $n_t = 92 + 6 = 98$, $n_d^2 = 130.3$, $n_d = 11.4$.

Inside diameter of shell, $d_s = P_T (n_d + 1) = 0.0603 \times 12.4\text{m} = 0.7487$ m. Nearest standard shell diameter is 29 inch, so $d_s = 0.7366$ m.

Revising the value of n_d; $n_d = (d_s/P_T) - 1 = (0.7366/0.0603) - 1 = 11.2$.

The number of rows of tubes accommodated in half the shell, $m = 0.577 n_d + 0.423 = (0.577 \times 11.2) + 0.423 = 6.894$, so maximum number of tubes, $n_t = m(2n_d - m) = 6.89(2 \times 11.2 - 6.89) = 107$.

Deducting a nozzle allowance of $\frac{1}{2}$ (11.2 +1) = 7 to ensure good flow patterns at the inlet and exit nozzles for the heating medium and also six tubes for positioning of tie rods, $n_t = 107 - (7 + 6)$. Number of tubes = 94 in a shell diameter of 0.737 m.

This implies that the mass velocity on the tube side is reduced below the initially assumed value of 3.588 kg/m²s and therefore it was recalculated and the new value used in the computer program for determination of reactor length.

Hence, $G_t = 3.588 \times 92/94 = 3.512$ kg/m² s.

4.7.2 Heat balance

$$Q = MC_{pm}(T_2 - T_1) + M\Delta H$$

where $T_1 = 773$K (Section 3.1) also let $T_2 = 663$K (reactor exit temperature) and for a fraction conversion of 0.9

$$M\Delta H = (0.9 \times 1557.2 \times 73900/74) \text{ kJ/h}$$

$$MC_{pm}(T_2 - T_1) = 1558.5 \times 6.706 \times (663-773) \text{ kJ/h}.$$

Heat required = $(-1.1532 + 1.3995) \times 10^6$ kJ/h, so $Q = 0.2464 \times 10^6$ kJ/h,

assuming $U_D = 0.05$ kW/m²K.

Heat transfer area available,

$$A = n_t \pi L \tfrac{1}{2}(d_i + d_o) \text{ m}^2$$
$$= 111 \times \pi \times 3.0 \times \tfrac{1}{2}(0.0409 + 0.0483) \text{ m}^2$$
$$= 46.65 \text{ m}^2.$$

Using $Q = U_D A \Delta T_{\ln}$: Logarithmic Mean Temperature Difference, $\Delta T_{\ln} = (0.2464 \times 10^6)/(0.050 \times 3600 \times 46.65)$ K = 29.3 K.

Assuming that the flue gas is available at 800 K, then the exit temperature, T_{fg2} to produce a ΔT_{ln} of 29.3 K is 750 K since $Q = RC_p'(T_{fg1} - T_{fg2})$ where specific heat of flue gas $C_p' = 1.195$ kJ/kg K at 775 K (Appendix G). Mass flowrate of flue gas, $R = (0.2464 \times 10^6)/(1.195 \times 50)$ kg/h = 4124 kg/h.

4.7.3 Tube side coefficient

Perry[19] lists three correlations that may be used to predict the wall transfer coefficient in a packed bed. Jakob's correlation was employed as the value of the coefficient calculated by this equation fell within the range of values expected for this reactor geometry and fluid properties. Hence

$$\frac{h_i d_t}{k_f} = f d_t^{0.17} \left(\frac{d_p G_t}{\mu_m}\right)^{0.83} \left(\frac{C_{pm}\mu_m}{k_f}\right)$$

where d_t is the inside tube diameter in feet. f is a function of the diameter ratio, d_p/d_t; now $d_p/d_t = 3.175/40.9 = 0.077$; from graph presented by Perry[19] $f = 0.195$.

$$\frac{h_i \times 0.0409}{0.102956} = 0.195 \times (0.1341)^{0.17}$$

$$\times \left(\frac{3.175 \times 10^{-3} \times 3.512}{1.877 \times 10^{-5}}\right)^{0.83}$$

$$\times \left(\frac{6.706 \times 10^3 \times 1.877 \times 10^{-5}}{0.102956}\right)$$

So $\quad h_i = 0.3488(594.0)^{0.83}(1.223)$ W/m²K $= 85.6$ W/m²K

Correcting this coefficient to the heat transfer area corresponding to the centre of the tube wall:

tube side coefficient $h_{iw} = h_i 2 d_i/(d_o + d_i)$

$= 85.6(2 \times 40.9)/(40.9 \times 48.3)$ W/m²K

$= 78.5$ W/m²K

4.7.4 Shell side coefficient

Calculation of the heat transfer coefficient for the shell side of a tubular exchanger using "j_H" factors is well established and is described by Kern[20]. In a recent publication, Butterworth[21] describes an alternative approach in which the heat transfer coefficient for flow over an "ideal" tube bundle of known configuration is calculated and then subsequently modified to account for by-pass, leakage and window effects caused by baffle design. This method will be adopted to calculate h_{ow}.

For heat transfer over an ideal tube bundle

$$Nu = a\, Re_c^m\, Pr^{0.34}\, F_N$$

Now, flow width per transverse pitch for a tube bank of the triangular (30°) type, $w = P_T - d_o = 0.0603 - 0.0483$ m $= 0.012$ m.

Minimum cross flow area, $S_m = L_w N_r$ where L is the exposed tube length = 0.250 m; N_r is the number of tubes per row. From Figure 10.1, $N_r = 9$, so $S_m = 0.250 \times 0.012 \times 9$ m² $= 0.027$ m²

Mass flowrate, $W = 4124/3600 = 1.1455$ kg/s.

Reynolds number, $Re_c = Wd_o/S_m \mu_{fg} = (1.1455 \times 0.0483)/(0.027 \times 3.4 \times 10^{-5}) = 60272$. For a staggered tube bank and $200\,000 > Re_c > 300$; $a = 0.273$, $m = 0.635$. Number of tube rows crossed = 11 (Figure 10.1). Then from Figure 7 (b)[21] $F_N = 1.01$.

$$\therefore Nu = 0.273\, Re_c^{0.635}\, Pr^{0.34} \times 1.01 = h_o' d_o/k_{fg}$$

$$\therefore h_o' = \frac{0.0465 \times 0.273 \times (60272)^{0.635} \times 1.01}{0.0483}$$

$$\times \left(\frac{1.195 \times 3.4 \times 10^{-5}}{0.0465 \times 10^{-3}}\right)^{0.34} \text{W/m}^2\text{K}$$

$$= 275.1 \text{ W/m}^2\text{K}$$

This value may now be corrected to compensate for by-pass, leakage and window effects using the following equation,

$$h_o = h_o' F_B F_L F_W$$

Using the dimensions stated in Figure 10.1 and the nomenclature of Figure 22[21], total area between tubes and around bundle, $S_M = \tfrac{1}{4}\pi(0.7366^2 - 102 \times 0.0483^2)$ m² $= 239.3 \times 10^{-3}$ m². Bypass flow area, $S_B = \tfrac{1}{4}\pi(0.7366 - 0.720^2)$ m² $= 18.99 \times 10^{-3}$ m².

$$\therefore S_B/S_M = 0.0793 \text{ and } Re_c > 100;$$

from Figure 23 (a)[21] $F_B = 0.89$

Considering the tube/baffle leakage; diameter of tube holes in baffles, $d_{tb} = 0.0491$ m, so $S_{tb} = \tfrac{1}{4}(102 \times \pi)(d_{tb}^2 - d_o^2) = 6.2422 \times 10^{-3}$ m².

Considering shell/baffle leakage d_s = 0.7366 m, d_B = 0.7318 m, so S_{sb} = $\frac{1}{4}\pi(d_s^2 - d_B^2)$ = 5.5357 × 10^{-3} m^2.

Total leakage flow area, $S_L = S_{tb} + S_{sb}$ = 11.778 × 10^{-3} m^2. Therefore S_L/S_M = 0.0492 and S_{sb}/S_L = 0.47 and from Figure 23 (b)[21] F_L = 0.88.

Total number of tubes (including tie rod spacers and tube containing reactor instrumentation) = 102. Number of tubes in two window zones = 30. Then ratio, R_w = 30/102 = 0.294, and from Figure 23 (c)[21], F_w = 1.06.

Then h_o = 275.1 × $F_B F_L F_W$ W/m^2K = 275.1 × 0.89 × 0.88 × 1.06 W/m^2K = 228.4 W/m^2K.

Correcting this value to the tube wall centre heat transfer area, shell side coefficient $h_{ow} = h_o (2d_o)/(d_o + d_i)$ = 228.4 (2 × 48.3)/(40.9 + 48.3) W/m^2 K = 247.3 W/m^2 K.

This heat transfer coefficient was also calculated using the "j_H" factor method as described in Section 4.10.4.3 and found to be 141.1 W/m^2K. Therefore the mean of these values will be employed in this design i.e. h_{ow} = 194.2 W/m^2 K.

Clean overall coefficient, $U_C = h_{iw} h_{ow}/(h_{iw} + h_{ow})$
= (78.5 × 194.2)/(78.5 + 194.2) W/m^2 K = 55.9 W/m^2 K.

Scale factors; for organic vapours, Rd_i = 0.0001 (Kern P845); for flue gas, Rd_o = 0.0004. Tube wall resistance, x/k_w = 3.68 × 10^{-3}/16.29 = 0.000226.

Overall design coefficient is given by

$$1/U_D = 1/U_C + Rd_i + Rd_o + x/k_w$$

$$= 1/55.9 + 0.0001 + 0.0004 + 0.000226 \ m^2K/W$$

$$\therefore U_D = 53.7 \ W/m^2K$$

This value compares favourably with the assumed value of 50 W/m^2K and therefore the heat exchange duty will be accomplished using 3.0 m tubes.

4.7.5 Determination of pressure drops

The tube side pressure drop was calculated using the equation presented by Hougen & Watson[13] for a randomly packed bed. Thus

$$\Delta P_t/Z = 2fG^2 v/\rho_m g \varepsilon^{1.7}$$

where $f = 2.60 (Re'')^{-0.3}$ for $150 > Re'' > 10$, and $f = 1.23 (Re'')^{-0.15}$ for $300 > Re'' > 150$, where $Re'' = G/v\mu_m$

Now G = 3.512 kg/m^2s, v = 1147 m^2/m^3, μ_m = 1.877 × 10^{-5} kg/m s, therefore, Re'' = 3.512/(1147 × 1.877 × 10^{-5}) = 163.

Therefore $f = 1.23 (163)^{-0.15} = 0.572$.

$$\therefore \Delta P_t = \frac{2 \times 0.572 \times 3.512^2 \times 1147 \times 3.0}{1.388 \times 9.81 \times (0.393)^{1.7}} \ kg/m^2$$

$$= 18539 \ kg/m^2$$

$$= 1.79 \ bar$$

The shell side pressure drop was calculated by the method outlined in Section 4.12.4.4. Hence $\Delta P_s = 0.40$ bar.

4.8 Computer output for reactor design

The important input parameters used in evaluation of the reactor length by the computer program are listed in Table 4.4. To determine the sensitivity of the values of the reactor length and exit conditions as provided by the program to the values of these parameters, the program was run for a range of the latter values. Although changes in all parameters affected the output to some extent, by far the most significant was the overall heat transfer coefficient between the flue gas and the fluid in the reactor tubes. This implies that any additional heat transfer resistance occurring such as excessive fouling or a large reduction in flowrate on either the shell or tube side, will produce a deterioration in reactor performance.

Table 4.4 – *Input parameters to computer program.*

FEED PRESSURE =	2.500	BAR
CATALYST ACTIVITY FACTOR =	0.770	
OVERALL HEAT TRANSFER COEFFICIENT =	0.050	KW/M**2 K
INLET FLUE GAS TEMPERATURE =	800.0	K

The final reactor length was 2.94 m which compares favourably with the tube length of 3.0 m which was found to be satisfactory in providing the necessary heat transfer as described in Section 4.7. There is, however, a significant difference when this value is compared to the value of 1.3 m as obtained using the "Height of a Reactor Unit" (HRU) approach in Section 4.2. This is attributable to the simplifications made in the latter by not considering heat transfer to the reactants, mass transfer effects, radial temperature variations and a decrease in catalyst activity, all of which have effects on the reactor design and should not be omitted. The exit stream conditions are summarised in Table 4.5.

Table 4.5 – *Output parameters from computer program.*

LENGTH OF REACTOR =	2.940	M
MEAN REACTOR TEMPERATURE =	653.8	K
MEAN PRODUCT TEMPERATURE =	643.0	K
FINAL REACTOR PRESSURE =	1.692	BAR
EXIT FLUE GAS TEMPERATURE =	779.2	K

The radial temperature profile is shown in Table 4.6 for reactor length increments of 5 cm and was then plotted in Figure 4.4 at intervals of 0.5 m. This shows a rapid fall in mean temperature as the conversion of 2-butanol occurs at the inlet and near the exit, where the rate of reaction decreases, the mean temperature rises due to an increased temperature difference across the tube wall. Due to the relatively high value of the effective diffusivity: velocity ratio, the radial conversion profile is flat as shown by Table 4.7. It may thus be concluded that a plug flow regime exists in the reactor tubes.

Table 4.6 — *Temperature profile.*

AXIS							WALL
773.0	773.0	773.0	773.0	773.0	773.0	773.0	773.0
723.9	723.9	723.9	723.9	723.9	724.0	724.1	726.9
704.9	704.9	704.9	704.9	704.9	705.0	705.7	711.7
693.6	693.6	693.6	693.6	693.7	693.8	695.2	704.0
685.6	685.6	685.6	685.6	685.6	685.9	688.1	699.4
679.1	679.1	679.1	679.1	679.2	679.7	682.6	696.1
673.9	673.9	673.9	673.9	674.0	674.7	678.6	694.0
669.5	669.5	669.5	669.5	669.7	670.7	675.4	692.6
665.7	665.7	665.8	665.8	666.0	667.3	672.9	691.6
662.5	662.5	662.5	662.5	662.9	664.5	670.9	691.0
659.5	659.5	659.5	659.6	660.1	662.0	669.3	690.6
656.9	656.9	656.9	657.0	657.6	659.9	668.0	690.4
654.5	654.5	654.5	654.7	655.4	658.1	667.0	690.3
652.3	652.3	652.3	652.5	653.3	656.4	666.1	690.3
650.3	650.3	650.3	650.6	651.5	655.0	665.4	690.4
648.4	648.4	648.4	648.7	649.9	653.8	664.9	690.5
646.6	646.6	646.7	647.1	648.4	652.7	664.4	690.7
645.0	645.0	645.1	645.5	647.0	651.7	664.1	690.9
643.4	643.5	643.6	644.1	645.8	650.9	663.9	691.2
642.0	642.0	642.2	642.7	644.6	650.1	663.7	691.5
640.6	640.6	640.8	641.5	643.6	649.4	663.6	691.8
639.3	639.4	639.6	640.3	642.6	648.9	663.6	692.1
638.1	638.1	638.4	639.2	641.7	648.4	663.6	692.4
636.9	637.0	637.2	638.2	640.9	647.9	663.6	692.7
635.8	635.9	636.2	637.2	640.2	647.6	663.7	693.0
634.7	634.8	635.2	636.3	639.5	647.2	663.8	693.4
633.7	633.8	634.2	635.5	638.8	647.0	663.9	693.7
632.8	632.9	633.3	634.7	638.3	646.8	664.1	694.0
631.8	632.0	632.4	633.9	637.8	646.6	664.3	694.3
631.0	631.1	631.6	633.2	637.3	646.4	664.5	694.7
630.1	630.3	630.8	632.6	636.8	646.3	664.7	695.0
629.3	629.5	630.1	631.9	636.5	646.3	665.0	695.3
628.5	628.7	629.4	631.4	636.1	646.2	665.2	695.6
627.8	628.0	628.7	630.8	635.8	646.2	665.5	695.9
627.1	627.3	628.1	630.3	635.5	646.2	665.7	696.2
626.4	626.6	627.5	629.9	635.3	646.3	666.0	696.6
625.8	626.0	626.9	629.4	635.0	646.3	666.3	696.8
625.1	625.4	626.4	629.0	634.9	646.4	666.6	697.1
624.5	624.8	625.9	628.6	634.7	646.5	666.9	697.4
624.0	624.3	625.4	628.3	634.6	646.6	667.2	697.7
623.4	623.8	624.9	628.0	634.4	646.7	667.5	698.0
622.9	623.3	624.5	627.7	634.4	646.9	667.8	698.3
622.4	622.8	624.1	627.4	634.3	647.0	668.1	698.5
621.9	622.3	623.7	627.2	634.2	647.2	668.5	698.8
621.5	621.9	623.3	626.9	634.2	647.4	668.8	699.0

621.1	621.5	623.0	626.7	634.2	647.6	669.1	699.3
620.7	621.1	622.7	626.6	634.2	647.8	669.4	699.5
620.3	620.8	622.4	626.4	634.2	648.0	669.7	699.8
619.9	620.4	622.1	626.3	634.3	648.2	670.1	700.0
619.6	620.1	621.9	626.1	634.3	648.5	670.4	700.3
619.2	619.8	621.6	626.0	634.4	648.7	670.7	700.5
618.9	619.5	621.4	626.0	634.5	649.0	671.0	700.7
618.6	619.2	621.2	625.9	634.6	649.2	671.4	700.9
618.4	619.0	621.0	625.8	634.7	649.5	671.7	701.1
618.1	618.7	620.9	625.8	634.8	649.8	672.0	701.3
617.9	618.5	620.7	625.8	635.0	650.0	672.3	701.6
617.7	618.3	620.6	625.8	635.1	650.3	672.7	701.7
617.5	618.2	620.5	625.8	635.3	650.6	673.0	701.9
617.3	618.0	620.4	625.8	635.4	650.9	673.3	702.1
617.1	617.8	620.3	625.8	635.6	651.2	673.6	702.3

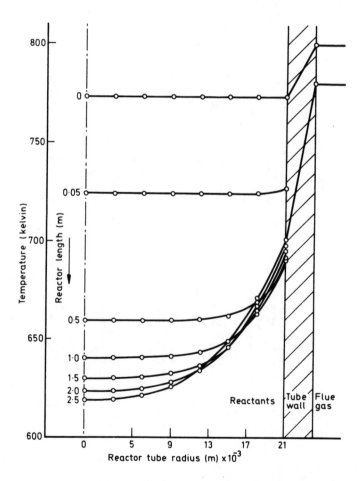

Figure 4.4 — *Temperature profile in reactor tubes.*

Table 4.7 — *Conversion profile.*

AXIS							WALL
0.0008	0.0008	0.0008	0.0008	0.0008	0.0008	0.0008	0.0008
0.1731	0.1731	0.1731	0.1731	0.1731	0.1731	0.1731	0.1731
0.2650	0.2650	0.2650	0.2650	0.2650	0.2650	0.2650	0.2650
0.3265	0.3265	0.3265	0.3265	0.3265	0.3265	0.3265	0.3265
0.3734	0.3734	0.3734	0.3734	0.3734	0.3734	0.3734	0.3734
0.4131	0.4131	0.4131	0.4131	0.4131	0.4131	0.4131	0.4131
0.4463	0.4463	0.4463	0.4463	0.4463	0.4463	0.4463	0.4463
0.4748	0.4748	0.4748	0.4748	0.4748	0.4748	0.4748	0.4748
0.4999	0.4999	0.4999	0.4999	0.4999	0.4999	0.4999	0.4999
0.5224	0.5224	0.5224	0.5224	0.5224	0.5224	0.5224	0.5224
0.5427	0.5427	0.5427	0.5427	0.5427	0.5427	0.5427	0.5427
0.5613	0.5613	0.5613	0.5613	0.5613	0.5613	0.5613	0.5613
0.5784	0.5784	0.5784	0.5784	0.5784	0.5784	0.5784	0.5784
0.5942	0.5942	0.5942	0.5942	0.5942	0.5942	0.5942	0.5942
0.6090	0.6090	0.6090	0.6090	0.6090	0.6090	0.6090	0.6090
0.6229	0.6229	0.6229	0.6229	0.6229	0.6229	0.6229	0.6229
0.6359	0.6359	0.6359	0.6359	0.6359	0.6359	0.6359	0.6359
0.6482	0.6482	0.6482	0.6482	0.6482	0.6482	0.6482	0.6482
0.6599	0.6599	0.6599	0.6599	0.6599	0.6599	0.6599	0.6599
0.6709	0.6709	0.6709	0.6709	0.6709	0.6709	0.6709	0.6709
0.6815	0.6815	0.6815	0.6815	0.6815	0.6815	0.6815	0.6815
0.6915	0.6915	0.6915	0.6915	0.6915	0.6915	0.6915	0.6915
0.7011	0.7011	0.7011	0.7011	0.7011	0.7011	0.7011	0.7011
0.7103	0.7103	0.7103	0.7103	0.7103	0.7103	0.7103	0.7103
0.7191	0.7191	0.7191	0.7191	0.7191	0.7191	0.7191	0.7191
0.7276	0.7276	0.7276	0.7276	0.7276	0.7276	0.7276	0.7276
0.7358	0.7358	0.7358	0.7358	0.7358	0.7358	0.7358	0.7358
0.7436	0.7436	0.7436	0.7436	0.7436	0.7436	0.7436	0.7436
0.7512	0.7512	0.7512	0.7512	0.7512	0.7512	0.7512	0.7512
0.7585	0.7585	0.7585	0.7585	0.7585	0.7585	0.7585	0.7585
0.7655	0.7655	0.7655	0.7655	0.7655	0.7655	0.7655	0.7655
0.7723	0.7723	0.7723	0.7723	0.7723	0.7723	0.7723	0.7723
0.7789	0.7789	0.7789	0.7789	0.7789	0.7789	0.7789	0.7789
0.7853	0.7853	0.7853	0.7853	0.7853	0.7853	0.7853	0.7853
0.7915	0.7915	0.7915	0.7915	0.7915	0.7915	0.7915	0.7915
0.7975	0.7975	0.7975	0.7975	0.7975	0.7975	0.7975	0.7975
0.8033	0.8033	0.8033	0.8033	0.8033	0.8033	0.8033	0.8033
0.8089	0.8089	0.8089	0.8089	0.8089	0.8089	0.8089	0.8089
0.8144	0.8144	0.8144	0.8144	0.8144	0.8144	0.8144	0.8144
0.8197	0.8197	0.8197	0.8197	0.8197	0.8197	0.8197	0.8197
0.8249	0.8249	0.8249	0.8249	0.8249	0.8249	0.8249	0.8249
0.8299	0.8299	0.8299	0.8299	0.8299	0.8299	0.8299	0.8299
0.8348	0.8348	0.8348	0.8348	0.8348	0.8348	0.8348	0.8348
0.8395	0.8395	0.8395	0.8395	0.8395	0.8395	0.8395	0.8395
0.8441	0.8441	0.8441	0.8441	0.8441	0.8441	0.8441	0.8441
0.8486	0.8486	0.8486	0.8486	0.8486	0.8486	0.8486	0.8486
0.8530	0.8530	0.8530	0.8530	0.8530	0.8530	0.8530	0.8530
0.8572	0.8572	0.8572	0.8572	0.8572	0.8572	0.8572	0.8572
0.8613	0.8613	0.8613	0.8613	0.8613	0.8613	0.8613	0.8613
0.8654	0.8654	0.8654	0.8654	0.8654	0.8654	0.8654	0.8654
0.8693	0.8693	0.8693	0.8693	0.8693	0.8693	0.8693	0.8693
0.8731	0.8731	0.8731	0.8731	0.8731	0.8731	0.8731	0.8731

```
0.8768  0.8768  0.8768  0.8768  0.8768  0.8768  0.8768  0.8768
0.8804  0.8804  0.8804  0.8804  0.8804  0.8804  0.8804  0.8804
0.8840  0.8840  0.8840  0.8840  0.8840  0.8840  0.8840  0.8840
0.8874  0.8874  0.8874  0.8874  0.8874  0.8874  0.8874  0.8874
0.8907  0.8907  0.8907  0.8907  0.8907  0.8907  0.8907  0.8907
0.8940  0.8940  0.8940  0.8940  0.8940  0.8940  0.8940  0.8940
0.8972  0.8972  0.8972  0.8972  0.8972  0.8972  0.8972  0.8972
0.9003  0.9003  0.9003  0.9003  0.9003  0.9003  0.9003  0.9003
```

DESIRED FRACTIONAL CONVERSION HAS BEEN ATTAINED

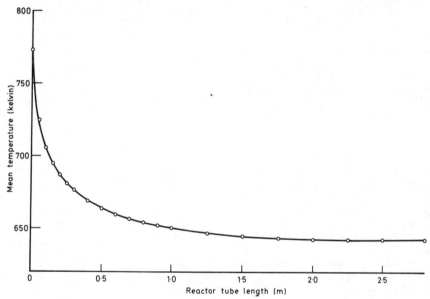

Figure 4.5 — *Variation of mean temperature along reactor tubes.*

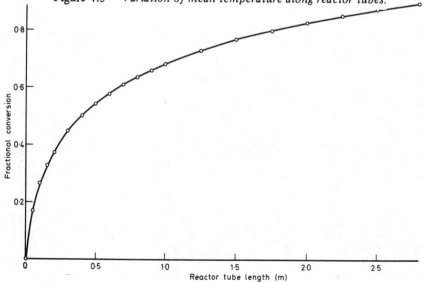

Figure 4.6 — *Variation of fractional conversion along reactor tubes.*

The mean values of fractional conversion, temperature and rate of reaction as listed in Table 4.8 are plotted against reactor tube length in Figures 4.5, 4.6 and 4.7.

Table 4.8 — *Variation of mean conversion, temperature and reaction rate with reactor length.*

REACTOR LENGTH (M)	MEAN FRACTIONAL CONVERSION	MEAN TEMPERATURE (DEG K)	RATE OF REACTION (KG.MOLE/M**2 S)
0.050	0.1731	724.1	0.1060E-03
0.100	0.2650	705.5	0.6219E-04
0.150	0.3265	694.6	0.4492E-04
0.200	0.3734	687.0	0.3544E-04
0.250	0.4131	680.9	0.3033E-04
0.300	0.4463	676.2	0.2570E-04
0.350	0.4748	672.3	0.2235E-04
0.400	0.4999	669.0	0.1981E-04
0.450	0.5224	666.2	0.1781E-04
0.500	0.5427	663.8	0.1619E-04
0.550	0.5612	661.7	0.1485E-04
0.600	0.5783	659.8	0.1371E-04
0.650	0.5942	658.2	0.1275E-04
0.700	0.6090	656.7	0.1191E-04
0.750	0.6228	655.3	0.1118E-04
0.800	0.6359	654.1	0.1053E-04
0.850	0.6482	653.0	0.9959E-05
0.900	0.6598	652.0	0.9444E-05
0.950	0.6709	651.1	0.8980E-05
1.000	0.6814	650.3	0.8558E-05
1.050	0.6915	649.5	0.8172E-05
1.100	0.7011	648.8	0.7817E-05
1.150	0.7103	648.1	0.7491E-05
1.200	0.7192	647.5	0.7189E-05
1.250	0.7276	647.0	0.6909E-05
1.300	0.7358	646.5	0.6649E-05
1.350	0.7437	646.0	0.6406E-05
1.400	0.7512	645.6	0.6179E-05
1.450	0.7585	645.2	0.5965E-05
1.500	0.7656	644.9	0.5765E-05
1.550	0.7724	644.6	0.5574E-05
1.600	0.7790	644.3	0.5393E-05
1.650	0.7854	644.0	0.5222E-05
1.700	0.7916	643.8	0.5060E-05
1.750	0.7976	643.5	0.4905E-05
1.800	0.8034	643.3	0.4758E-05
1.850	0.8091	643.2	0.4618E-05
1.900	0.8146	643.0	0.4483E-05
1.950	0.8199	642.9	0.4355E-05
2.000	0.8251	642.8	0.4232E-05
2.050	0.8301	642.6	0.4113E-05
2.100	0.8350	642.6	0.3999E-05
2.150	0.8397	642.5	0.3888E-05
2.200	0.8444	642.4	0.3782E-05
2.250	0.8488	642.4	0.3680E-05
2.300	0.8532	642.4	0.3581E-05
2.350	0.8575	642.3	0.3485E-05
2.400	0.8616	642.3	0.3393E-05
2.450	0.8657	642.3	0.3304E-05
2.500	0.8696	642.4	0.3218E-05
2.550	0.8734	642.4	0.3133E-05
2.600	0.8771	642.4	0.3051E-05
2.650	0.8808	642.5	0.2971E-05
2.700	0.8843	642.5	0.2894E-05
2.750	0.8877	642.6	0.2819E-05
2.800	0.8911	642.7	0.2746E-05
2.850	0.8944	642.8	0.2674E-05
2.900	0.8975	642.9	0.2605E-05

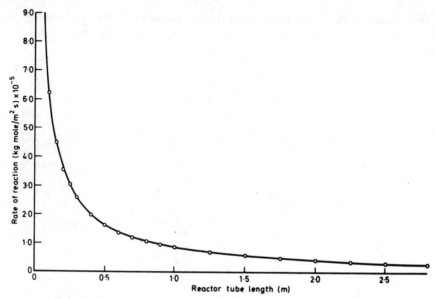

Figure 4.7 – *Reaction rate variation with reactor tube length.*

Summaries of the reaction rate calculation are shown for low, intermediate and high values of fractional conversion in Tables 4.9, 4.10 and 4.11 respectively. They confirm that significant differences do exist between interfactial and bulk phase properties (*viz* partial pressures and temperatures) especially at low fractional conversion values.

Table 4.9 – *Reactor and fluid properties at low conversion stage of 2-butanol.*

```
                **** SUMMARY OF REACTION RATE CALCULATION ****

                    MEAN TEMPERATURE   =   737.3    DEG K
                     (IN BULK PHASE)

                    MEAN TEMPERATURE   =   724.5    DEG K
                     (AT INTERFACE)

                    MEAN FRACTIONAL    =   0.118
                    CONVERSION

                    PRESSURE IN REACTOR= 2.4948    BAR
```

**** PROPERTIES OF COMPONENT VAPOUR PHASES ****

	2-BUTANONE	2-BUTANOL	HYDROGEN	
DENSITY	2.9954	3.0786	0.0832	KG/M**3
SPECIFIC HEAT	2.5871	2.8313	14.7648	KJ/KG DEG K
VISCOSITY	0.1843E-04	0.1999E-04	0.1609E-04	KG/M S
THERMAL CONDUCTIVITY	0.0642	0.0761	0.3563	W/M DEG K.
PRANDTL NUMBER	0.7426	0.7438	0.6902	
DIFFUSIVITY	0.9568E-05	0.2221E-04	0.4334E-04	M**2/S

SCHMIDT NUMBER	0.6431	0.2923	4.4634	
PARTIAL PRESSURE (IN BULK PHASE)	0.2627	1.9693	0.2627	BAR
PARTIAL PRESSURE (AT INTERFACE)	0.3297	1.8896	0.4969	BAR

HEAT OF REACTION = 73884.3 KJ/KG MOLF

RATE OF REACTION = 0.1598E-03 KG MOLE/M**2 S

Table 4.10 — *Reactor and fluid properties at intermediate conversion stage of 2-butanol.*

**** SUMMARY OF REACTION RATE CALCULATION ****

MEAN TEMPERATURE (IN BULK PHASE) = 670.3 DEG K

MEAN TEMPERATURE (AT INTERFACE) = 669.3 DEG K

MEAN FRACTIONAL CONVERSION = 0.490

PRESSURE IN REACTOR = 2.4080 BAR

**** PROPERTIES OF COMPONENT VAPOUR PHASES ****

	2-BUTANONE	2-BUTANOL	HYDROGEN	
DENSITY	3.1526	3.2402	0.0876	KG/M**3
SPECIFIC HEAT	2.4473	2.6663	14.7047	KJ/KG DEG K
VISCOSITY	0.1690E-04	0.1833E-04	0.1536E-04	KG/M S
THERMAL CONDUCTIVITY	0.0558	0.0658	0.3273	W/M DEG K.
PRANDTL NUMBER	0.7420	0.7432	0.6900	
DIFFUSIVITY	0.1993E-04	0.1951E-04	0.3843E-04	M**2/S
SCHMIDT NUMBER	0.2690	0.2900	4.5635	
PARTIAL PRESSURE (IN BULK PHASE)	0.7922	0.8237	0.7922	BAR
PARTIAL PRESSURE (AT INTERFACE)	0.7957	0.8162	0.8157	BAR

HEAT OF REACTION = 72211.6 KJ/KG MOLF

RATE OF REACTION = 0.2647E-04 KG MOLE/M**2 S

Table 4.11 – *Reactor and fluid properties at high conversion stage of 2-butanol.*

```
**** SUMMARY OF REACTION RATE CALCULATION ****

        MEAN TEMPERATURE      =   642.3   DEG K
        (IN BULK PHASE)

        MEAN TEMPERATURE      =   642.2   DEG K
        (AT INTERFACE)

        MEAN FRACTIONAL       =   0.880
        CONVERSION

        PRESSURE IN REACTOR   =   1.7872  BAR
```

**** PROPERTIES OF COMPONENT VAPOUR PHASES ****

	2-BUTANONE	2-BUTANOL	HYDROGEN	
DENSITY	2.4415	2.5093	0.0678	KG/M**3
SPECIFIC HEAT	2.3791	2.5858	14.6795	KJ/KG DEG K
VISCOSITY	0.1620E-04	0.1757E-04	0.1501E-04	KG/M S
THERMAL CONDUCTIVITY	0.0520	0.0612	0.3193	W/M DEG K.
PRANDTL NUMBER	0.7417	0.7429	0.6899	
DIFFUSIVITY	0.2377E-04	0.1830E-04	0.3622E-04	M**2/S
SCHMIDT NUMBER	0.2792	0.3825	6.1099	
PARTIAL PRESSURE (IN BULK PHASE)	0.8368	0.1136	0.8368	BAR
PARTIAL PRESSURE (AT INTERFACE)	0.8370	0.1130	0.8388	BAR

```
        HEAT OF REACTION   =  71384.8  KJ/KG MOLE
        RATE OF REACTION   = 0.3851E-05 KG MOLE/M**2 S
```

Concluding, a tubular catalytic reactor containing 94 tubes of 40.9 mm diameter and 3.0 m in length will be adequate to achieve 90% conversion of 2-butanol to 2-butanone and hydrogen whilst providing sufficient area for the necessary heat transfer.

4.9 Catalyst regeneration and production scheduling

As the catalyst requires regeneration and reduction after 15 h[13] three reactors are required which are subject to a cycle of 43.5 h. After the reaction time of 15 h, the reactor is to be taken off-line and air at 700 K passed through the tubes for 5 h. Hydrogen at the same temperature for the same time is then passed through the reactor to effect reduction. Also to eliminate the risk of explosion, it is necessary to purge the reactor with nitrogen for half an hour between the oxidation and reduction stages of regeneration. It appears that, as the regeneration time is less than the on-line time, two reactors would be used for safety, bearing in mind the times required for stream changeover and allowing for any malfunction in the reactors.

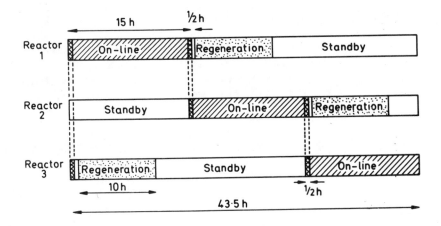

Figure 4.8 — *Reactor utilisation schedule.*

The scheduling for regeneration is illustrated in Figure 4.8. The important features of the schedule are firstly that before a reactor is taken off-line, the succeeding reactor is brought on stream a half hour before to ensure continuous production. Another half hour period of stand-by is allocated before commencement of the regeneration procedure. Secondly, it should be observed that at any time during the cycle, at least one reactor is available on stand-by and at some stages a choice of reactors is available, thus facilitating flexibility of operation. Also if anything occurs to halt operation in the MEK recovery plant, the alcohol feed can be put through the reactors in parallel thus effecting a higher conversion and minimising product loss.

4.10 Design of feed heating equipment

4.10.1 *General discussion*

The heating of the cold 2-butanol feed is to be completed in three basic stages in order to avoid large heat losses. The cold feed is firstly preheated to its boiling point using steam as the heating medium and then vaporised in a thermo-syphon reboiler utilising the heat contained in the reaction products. The vapour, which has had most of the entrained liquid removed in a knock-out drum, is then heated to reaction temperature employing exhaust flue gas from the reactor. This arrangement, besides utilising otherwise wasted heat resources, was also considered to be favourable in terms of temperature control and access for maintenance purposes.

4.10.2. *Cold feed preheater*

1558.5 kg/h of 2-butanol is to be heated from 288 K, the storage temperature, to its boiling point at 373 K. The detailed design of the heat exchanger is not presented here as this will be more fully described in Section 4.10.4.1.

Consider a heat balance, $Q = M C_p (T_1 - T_2)$ where $C_p = 1.497$ kJ/kgK at $\frac{1}{2}(T_1 + T_2) = 330.5$ K (see Appendix H). $Q = 1558.5 \times 1.497 (373-288)$ kJ/h = 0.1983×10^6 kJ/h.

The heating medium to be used is dry saturated steam at 413 K then Enthalpy, h_{fg} = 2145 kJ/kg, therefore steam flowrate, $s = 0.1983 \times 10^6/2145 = 92.4$ kg/h

For an estimation of the overall design coefficient, U_D, estimates of the individual heat transfer resistances will be added, thus[10] $h_{io} = 8500$ W/m²K, $h_o = 1100$ W/m² K.

It should be noted that steam is placed in tubes to alleviate corrosion of both shell and tube sides due to the condensate produced during heat transfer. For industrial organic liquids $R_{do} = 0.00017$ (Kern P845). For steam $R_{di} = 0.0$, so,

$$1/U_D = 1/h_{io} + 1/h_o + R_{do} + R_{di} = \frac{1}{8500} + \frac{1}{1100} + 0.00017 = 1.203 \times 10^{-3} \, m^2 K/W$$

$$\therefore U_D = 831 \; W/m^2 K$$

Log Mean Temperature Difference is given by

$$\Delta T_{ln} = \frac{(T_s - T_2) - (T_s - T_1)}{\ln[(T_s - T_2)/(T_s - T_1)]}$$
$$= \frac{(413 - 288) - (413 - 373)}{\ln[(413 - 288)/(413 - 373)]} = 74.6 \; K$$

Now, $Q = U_D A \, \Delta T_{ln}$, therefore $A = Q/U_D \, \Delta T_{ln}$
$= (0.1983 \times 10^6)/(8.31 \times 74.6) \, m^2 = 3.2 \, m^2$.

If ¾ inch nominal size tubes of 4ft length are employed o.d. = 26.7 mm; i.d. = 21.0 mm [BS 1600(ii)]. Surface area per unit length = 0.0748 m²/m, therefore surface area per tube = 0.09119 m² and number of tubes required = 3.2/0.09119 = 36.

Therefore a single pass shell and tube exchanger comprising 36 tubes on a 24 mm triangular pitch within a 203 mm i.d. shell will be adequate for the specified duty. The pressure drop across the 2-butanol preheater was calculated using the equation presented in Section 4.11.4.4 to be negligible *i.e.* 10^{-4} bar.

4.10.3 *2-butanol vaporiser*

The energy requirement for vaporising the 2-butanol feed at 373 K is similar to the heat available when cooling the reaction products from 642 K to 398 K, the condenser inlet temperature. The driving force for heat transfer is also satisfactory and therefore this arrangement for heat recovery is acceptable. However, during the start-up period, no heat is available from the reaction products and it is necessary to provide steam heating facilities with modification to the process route to prevent contamination of the MEK product by steam condensate. This is discussed in further detail in Section 12.5. Since the heat transfer coefficient associated with condensing steam is far greater than that of cooling the reaction products, the vaporiser will be designed at steady state employing the latter as the heat source to ensure satisfactory performance for all operating conditions.

Relatively high transfer coefficients may be realised when boiling occurs in tubes and a vertical tube boiler was selected. Also to minimise capital and pumping costs, a thermosyphon type of vaporiser will be employed because the relatively pure feed is not viscous and fouling is not likely to be excessive[23].

This design follows the basic procedure outlined by Holland et al[18] for vaporisation of an organic liquid in a vertical thermo-syphon reboiler.

4.10.3.1 *Heat transfer area*
From the definition of the overall heat transfer coefficient,

$$\frac{1}{U_D} = \frac{1}{h_i} + \frac{x_w}{k_w} + \frac{1}{h_o} + Rd_o + Rd_i$$

from Table P9.1[18] for C_2–C_7 alcohols

$$\left(\frac{x_w}{k_w} + Rd_i + \frac{1}{h_i}\right)^{-1} = 1420 \text{ W/m}^2\text{K}$$

also assuming shell side coefficient, $h_o = 170 \text{ W/m}^2\text{K}$, $Rd_o = 0.0001 \text{ m}^2\text{K/W}$

so, $1/U_D = 1/1420 + 1/170 + 0.0001 \text{ m}^2 \text{ K/W}$

$$\therefore U_D = 149.7 \text{ W/m}^2\text{K}$$

The reaction product vapours enter at 642 K cooling to 398 K whilst the alcohol boils at 373 K. Therefore mean temperature difference,

$$\Delta T_{av} = \tfrac{1}{2}[(642-373) + (398-373)] \text{ K} = 151 \text{ K}.$$

The heat flux required is given by, $(Q/A) = U_D \Delta T_{av} = 149.7 \times 151 \text{ W/m}^2 = 22\,604 \text{ W/m}^2$.

Kern[20] recommends a maximum flux of 38 000 W/m² for the vaporisation of organic liquids. The actual flux represents 60% of the maximum which is acceptable for design purposes.

Heat load on vaporiser, $Q = m\lambda = 1558.5 \times 562.2 \text{ kJ/h} = 0.8762 \times 10^6 \text{ kJ/h}$. Heat supplied by reaction products, $Q = 1558.5 \times 2.304 \times (642-398) = 0.8762 \times 10^6 \text{ kJ/h}$.

Heat transfer area required, $A = Q/(Q/A) = (0.8762 \times 10^6)/(3600 \times 22\,604) \text{ m}^2 = 10.76 \text{ m}^2$.

Assuming a single tube pass, $\tfrac{3}{4}$ inch nominal size tubes of length 2.5 m, inside diameter 20.96 mm and outside diameter 26.7 mm were selected. Using the procedure described in Section 4.10.4.1, 59 tubes in a 0.3302 m diameter shell provides the required heat transfer area.

4.10.3.2 *Length of sensible heating section*
Although the 2–butanol feed enters the vaporiser at 373 K boiling is suppressed at the bottom of the tubes due to the hydrostatic head of liquid and vapour. Consequently, it is necessary to calculate the length of tube required for sensible heating of the liquid to its boiling point. This section of tube is represented on Figure 4.11 by L_H.

Since the rate of heat transfer to the liquid depends on flowrate, it is necessary to assume a value of the fraction of the liquid that is vaporised in the tubes. Initially, let the fractional exit vaporisation, $w_e = 0.15$.

Liquid mass flowrate, $M_T = 1558.5/(3600 \times 0.15) = 2.886$ kg/s

Tubeside flow area, $S_i = \frac{1}{4} n_T \pi d_i^2 = \frac{1}{4}(59 \times \pi \times 0.02096^2)$ m² $= 0.02035$ m²

Mass velocity, $G_{Li} = M_T/S_i = 2.886/0.02035$ kg/m²s $= 141.8$ kg/m²s

Reynolds Number, $Re_{Li} = G_{Li} d_i/\mu_L$
(where $\mu_L = 0.231 \times 10^{-3}$ kg/m s, Appendix H)
$= (141.8 \times 0.02096)/(0.231 \times 10^{-3}) = 12\,866$

Prandtl Number, $Pr_{Li} = C_{pL}\mu_L/k_L$
(where $k_L = 0.1817$ W/m K, and $C_{pL} = 3.60$ kJ/kg K, Appendix H)
$= (3.60 \times 0.231 \times 10^{-3})/(0.1817 \times 10^{-3}) = 4.576$.

For turbulent flow in pipes, the inside heat transfer coefficient is given by

$$Nu_L = h_{Li}d_i/k_L = 0.023(Re_{Li})^{0.8}(Pr_{Li})^{0.4}.$$

Therefore heat transfer coefficient for the liquid in the sensible heating section is,

$$h_{Li} = 0.023(0.1817/0.02096)(12866)^{0.8}(4.576)^{0.4} \text{ W/m}^2 \text{ K}$$
$$= 710 \text{ W/m}^2\text{K}$$

Now $\quad Q = \dfrac{A(T_w - T)}{(1/h_i) + Rd_i + (x_w/k_w)} \quad$ where T_w is the wall temperature

$$= \dfrac{A\Delta T_{av}}{(1/h_i) + Rd_i + (x_w/k_w) + Rd_o + (1/h_o)}$$

by equating the above expressions,

$$\dfrac{T_w - T}{\Delta T_{av}} = \dfrac{(1/h_i) + Rd_i + (x_w/k_w)}{(1/h_i) + Rd_i + (x_w/k_w) + Rd_o + (1/h_o)}$$

$$= \dfrac{0.0007}{0.0007 + 0.00009 + 0.00588}$$

$$= 0.105$$

From the design value of $\Delta T_{av} = 151$ K, $T_w - T = 15.8$ K.
Considering a heat balance over a tube length of ΔL

$$h_{Li}(n_T \pi d_i \Delta L)(T_w - T) = M_T C_{pL} \Delta T$$

∴ Temperature gradient $\dfrac{\Delta T}{\Delta L} = \dfrac{710.0\,(59 \times \pi \times 0.02096) \times 15.8}{2.886 \times 3.60 \times 10^3}$ K/m $= 4.20$ K/m.

Pressure gradient $(-\Delta P/\Delta L) = \rho_L G$
(where $\rho_L = 709$ kg/m³, Appendix H)
$= 709 \times 9.81$ N/m² $= 6955$ N/m².

Also from Figure 4.9 $(\Delta T/\Delta P) = 3.25 \times 10^{-4}$ Km²/N.

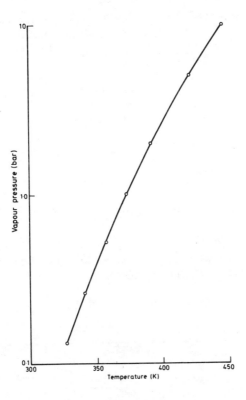

Figure 4.9 — *Vapour pressure variation with temperature for 2-butanol*[19].

The length of the sensible heating section is given by Holland et al[18] as,

$$\frac{L_H}{L_H + L_V} = \left(\frac{\Delta T}{\Delta P}\right)_s \bigg/ \left\{\left[\left(\frac{\Delta T}{\Delta L}\right) \bigg/ \left(-\frac{\Delta P}{\Delta L}\right)\right] + \left(\frac{\Delta T}{\Delta P}\right)_s\right\}$$

$$= \frac{3.25 \times 10^{-4}}{(4.20/6955) + (3.25 \times 10^{-4})}$$

$$= 0.350$$

The total length of tubes, $L_H + L_V$ = 2.5 m. Length of sensible heating section, L_H = 0.875 m.

4.10.3.3 *Calculation of circulation rate*

The fractional vaporisation was assumed to be 0.15 in Section 4.10.3.2 and it is essential that the associated circulation rate in the vaporiser is achieved. The circulation rate depends on the sum of the resistances to flow in the inlet pipe from the knock-out drum to the bottom of the tubes, the vaporiser tubes and the exit pipe from the top of the tubes to the knock out drum.

For the inlet pipe, liquid flowrate, $M_{TI} = 2.453$ kg/s. Let diameter of inlet pipe, $D_I = 0.05248$ m (2 inch nominal size) which produces a pipe velocity of approximately 1.5 m/s. Flow area, $S_I = 2.163 \times 10^{-3}$ m^2. $G_{LI} = M_{TI}/S_I = 2.453/(2.163 \times 10^{-3})$ kg/m^2 s = 1134 kg/m^2 s.

Reynolds Number, $Re_{LI} = (1134 \times 0.05248)/(0.231 \times 10^{-3}) = 257\,630$

For commercial tubes, friction factor is given by

$$j_f = 0.02117\,(Re)^{-0.164} \text{ for } 1000 < Re < 100000$$

$$j_f = 0.05573\,(Re)^{-0.261}\; Re > 100000$$

Hence friction factor, $(j_f)_{LI} = 0.05573\,(257\,630)^{-0.261} = 0.00216$

For the exit pipe, assuming a pipe diameter of $3\frac{1}{2}$ inch nominal size, $D_E = 0.09012$ m; similarly, $S_E = 6.379 \times 10^{-3}$ m^2, $G_{LE} = 384.6$ kg/m^2 s, $Re_{LE} = 150000$, therefore $(j_f)_{LE} = 0.00248$.

For the vaporiser tubes, from Section 4.10.3.2., $Re_{Li} = 12866$, hence $(j_f)_{Li} = 0.00448$.

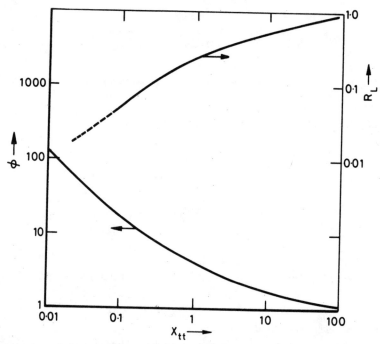

Figure 4.10 — *Relationship of Lockhart and Martinelli parameters for two phase flow.* [Based on Figure G-16, *Heat Transfer* by F.A. Holland *et al*, by permission of the publishers, Heinemann Educational Books]

To evaluate the two phase density, ρ_{TP} existing in the vaporiser tubes, and the acceleration loss term for the thermo-syphon leg, use was made of the Lockhart and Martinelli parameters for two phase flow which are presented in graphical form in Figure 4.10. The Lockhart and Martinelli parameter is defined as,

$$X_{tt} = \left[\frac{(1-w)}{w}\right]^{0.9} \left[\frac{\rho_v}{\rho_L}\right]^{0.5} \left[\frac{\mu_L}{\mu_v}\right]^{0.1}$$

where $\rho_v = 2.971$ kg/m³ (Appendix F), $\mu_v = 1.02 \times 10^{-5}$ kg/m s (Appendix A), $w = w_e = 0.15$.

$$\therefore X_{tt} = \left[\frac{(1-0.15)}{0.15}\right]^{0.9} \left[\frac{2.971}{709}\right]^{0.5} \left[\frac{0.231 \times 10^{-3}}{1.020 \times 10^{-5}}\right]^{0.1}$$

$$= 0.421$$

From Figure 4.10, volume fraction of liquid phase, $R_L = 0.14$. Parameter for two phase pressure loss, $\phi_e = 6.7$. Then two phase density, ρ_{TP} is given by

$$\rho_{TP} = \rho_v R_v + \rho_L R_L$$
$$= (2.971 \times 0.86) + (709 \times 0.14) \text{ kg/m}^3$$
$$= 101.8 \text{ kg/m}^3$$

Acceleration loss group is

$$\gamma = \left[\frac{(1-w)^2}{R_L}\right] + \left[\frac{\rho_L w^2}{\rho_v R_v}\right] - 1$$

For $w = w_e$: $\gamma_e = (0.85^2/0.14) + [(709 \times 0.15^2)/(2.971 \times 0.86)] - 1 = 10.4$.
Also for $w = \frac{1}{3}w_e$: $X_{tt} = 1.252$. From Figure 4.10, $R_L = 0.25$, then $(\rho_{TP})_m = 179.5$ kg/m³ and for $w_m = 2w_e/3$: $X_{tt} = 0.639$, then $\phi_m^2 = 29.2$.

The circulation rate is given by Holland et al[18] as

$$M_T^2 = \rho_L g S_I^2 L_v [\rho_L - (\rho_{TP})_m]/\psi$$

where
$$\psi = \gamma_e (S_I/S_E)^2 + 4(d_f)_{LI}(L_I/D_I)$$
$$+ 4(d_f)_{Li}(S_I/S_i)^2[(L_H/d_i) + \phi_m^2(1-w_m)^2(L_v/d_i)]$$
$$+ 4(d_f)_{LE}(1-w_e)^2(S_I/S_E)^2 \phi_e^2 (L_E/D_E)$$

$$= 10.4\left(\frac{2.163 \times 10^{-3}}{6.379 \times 10^{-3}}\right) + 4 \times 0.00216\left(\frac{4.0}{0.05248}\right)$$

$$+ 4 \times 0.00448\left(\frac{2.163 \times 10^{-3}}{0.02035}\right) + \left[\left(\frac{0.875}{0.02096}\right) + 29.2(0.9)^2\left(\frac{1.625}{0.02096}\right)\right]$$

$$+4 \times 0.00248(0.85)^2 \left(\frac{2.163 \times 10^{-3}}{6.379 \times 10^{-3}}\right)^2 44.9 \left(\frac{1.5}{0.09012}\right)$$

$$= 1.195 + 0.659 + 0.380 + 0.616$$

$$= 2.850$$

Thus $M_T{}^2 = [709 \times 9.81 \times (2.163 \times 10^{-3})^2 \times 1.625 \,(709-179.5)]/2.850 \text{ kg}^2/\text{s}^2$
$= 9.82 \text{ kg}^2/\text{s}^2$, and the circulation rate, $M_T = 3.133$ kg/s.

This value represents a difference of 8% based upon the circulation rate associated with 0.15 fractional vaporisation and therefore another trial is necessary.

The calculations in Sections 4.10.3.2 and 4.10.3.3 were repeated with $w_è = 0.14$ in a second trial to produce a circulation rate, $M_T = 3.152$ kg/s which differs by 2% from the value of $M_T (3.092 \text{kg/s})$ corresponding to 14% fractional vaporisation.

4.10.3.4 *Pressure drop across the sensible heating section*

The pressure at the top of the sensible heating section is,

$$P_C = P_A + \rho_L g L_v - (\Delta P_f)_{AB} - (\Delta P_f)_{BC}.$$

Friction loss in inlet pipe, $(\Delta P_f)_{AB}$
$= 4 \, (j_f)_{LI} \, (L_I/D_I) \, G_{LI}{}^2 / \rho_L = 4 \times 0.00211 \,(4.0/0.05248)\, 1229^2/709 \text{ N/m}^2$
$= 1370 \text{ N/m}^2$.

Friction loss in tubes, $(\Delta P_f)_{BC}$
$= 4 \, (j_f)_{Li} \, (L_H/d_i) \, G_{Li}{}^2/\rho_L$
$= 4 \times 0.00443 \,(0.882/0.02096)\, 151.9^2/709 \text{ N/m}^2$
$= 24 \text{ N/m}^2$.

As the 2-butanol feed enters at atmospheric pressure $P_A = 101\,323 \text{ N/m}^2$. Static head provided by the thermo-syphon leg, $\rho_L g L_v = 709 \times 9.81 \times 1.618$ $= 11\,253 \text{ N/m}^2$. Therefore $P_C = (101\,323 + 11\,253 - 1370 - 24) \text{ N/m}^2 = 111\,182$ N/m² $= 1.097$ bar.

From Figure 4.9 the boiling temperature at this pressure is 375 K and it will not be necessary to recalculate the physical properties since they were evaluated at 373 K.

Static pressure loss in sensible heating section is $\rho_L g L_H = 709 \times 9.81 \times 0.882 \text{ N/m}^2 = 6134 \text{ N/m}^2$. Total Pressure loss, $\Delta P_{BC} = (\Delta P_f)_{BC} + \rho_L g L_H$ $= 24 + 6134 \text{ N/m}^2 = 6158 \text{ N/m}^2$.

4.10.3.5 *Length of vaporising section*

The heat transfer rate to the boiling liquid is dependent upon the Reynolds Number in the tubes and because a phase change occurs the value of Re changes rapidly. It is therefore convenient to calculate the boiling heat transfer coefficient for a number of increments in this section. Three increments were selected based upon intervals of equal vaporisation of 4.666%. The Lockhart and Martinelli parameters for two phase flow were evaluated using the procedure described in Section 4.10.3.3 for the inlet, mid-point and exit of each increment and the results summarised in Table 4.12

Table 4.12 – *Two phase properties in vaporising section increments*

Increment	Fractional Vaporisation w	Lockhart Martinelli Parameter X_{tt}	Volume Fraction Liquid R_L	Pressure Drop Factor ϕ^2	Two Phase Density ρ_{TP}
ΔL_{12}	0.02333	2.548	0.33	7.8	235.9
	0.04666	1.336	0.27	12.3	193.6
ΔL_{23}	0.0700	0.907	0.21	18.5	151.2
	0.09333	0.684	0.19	25.0	137.1
ΔL_{34}	0.11666	0.547	0.17	36.0	123.0
	0.1400	0.453	0.14	42.3	101.8

For the first increment, ΔL_{12}, mass flowrate, $M_T = 3.092$ kg/s, flow area, $S_i = 0.02035$ m², velocity at inlet of increment, $u_1 = M_T/[(\rho_{TP})_1 S_i] = 0.644$ m/s, similarly, velocity at outlet of increment, $u_2 = 0.785$ m/s.

Logarithmic mean velocity, $u_m = (u_2 - u_1)/\ln(u_2/u_1)$
$= (0.785 - 0.644)/\ln(0.785/0.644)$
$= 0.712$ m/s.

Reynolds Number, $Re_m = \rho_L u_m d_i/\mu_L = 709 \times 0.712 \times 0.02096/(0.231 \times 10^{-3}) = 45815$.

As calculated in Section 4.10.3.2, $Pr_L = 4.576$. The boiling film heat transfer coefficient, h_{vi} was determined using Hughmark's correlation[24]

$$Nu_{vi} = \frac{h_{vi} d_i}{k_L} = \frac{204 Pr_L^{0.25} Re_m^{0.55}}{[(\rho_L - \rho_v)/\rho_v]^{0.8} [1/\sigma C_{pL}]^{0.45} [(L_v + L_H)/10]^{0.25}}$$

Since this correlation is empirical, it was necessary to evaluate the latter two terms of the denominator in the f.p.s. system of units. Thus, surface tension, $= 15.15 \times 10^{-3}$ N/m (Appendix H) $= 0.334$ pdl/ft, liquid specific heat, $C_{pL} = 0.86$ Btu/lb deg F, total tube length, $(L_V + L_H) = 8.202$ ft. Therefore the boiling film heat transfer coefficient is

$$h_{vi} = \frac{0.1817 \times 204 \times (4.576)^{0.25} \times (45815)^{0.55}}{[(709 - 2.971)/2.971]^{0.8} [1/(0.86 \times 0.0334)]^{0.45} [8.202/10]^{0.25}} \text{ W/m}^2\text{K}$$

$= 2528$ W/m²K

Considering a heat balance over the first increment

$$h_{vi}(n_T \pi d_i \Delta L_{12})(T_w - T) = M_T \lambda \Delta w_{12}$$

where the heat of vaporisation, $\lambda = 562.2$ kJ/kg, fractional vaporisation in increment, $\Delta w_{12} = 0.04666$, then length of increment:

$$\Delta L_{12} = \frac{(3.092 \times 562.2 \times 0.04666 \times 10^3)}{(15.8 \times 2528 \times 59 \times \pi \times 0.02096)} \text{ m} = 0.523 \text{ m}$$

Table 4.13 – *Calculation of vaporisation section increments*

Calculation Sequence		ΔL_{12}	ΔL_{23}	ΔL_{34}	Σ
u_1	(m/s)	0.644	0.785	1.108	
u_2	(m/s)	0.785	1.108	1.493	
u_m	(m/s)	0.712	0.937	1.292	
Re_m	(-)	45815	60269	83158	
h_{vi}	(W/m² K)	2528	2940	3509	
ΔL	(m)	0.523	0.450	0.377	1.350
$g(\rho_{TP})_m \Delta L$	(N/m²)	1210	667	455	
ΔP_f	(N/m²)	108	201	299	608

Similar calculations were performed for the second and third increments, ΔL_{23} and ΔL_{34} and the results summarised in Table 4.13. The total tube length required is then given by

$$L_v + L_H = \Delta L_{12} + \Delta L_{23} + \Delta L_{34} + L_H =$$
$$0.523 + 0.450 + 0.377 + 0.882 = 2.23 \text{ m}$$

Therefore the initially assumed length of 2.5 m will be satisfactory for the duty proposed.

4.10.3.6. *Pressure drop across vaporising section*

A preliminary estimate of the resistance to flow in the tubes was made in Section 4.10.3.3 based upon mean properties of the liquid and vapour. However, more accurate analysis of the pressure drop in the vaporising section is now possible since the relative proportions of liquid and vapour in each increment are known.

Considering the first increment, ΔL_{12}, static pressure loss = $g(\rho_{TP})_{am} \Delta L_{12}$ [where $(\rho_{TP})_{am}$ is the arithmetic mean two phase density as presented in Table 4.12] = 9.81 × 235.9 × 0.523 N/m² = 1210 N/m²

Frictional pressure loss,

$$(\Delta P_f)_{12} = 4(j_f)_{Li}(\Delta L_{12}/d_i)\phi_{am}^2 G_{TLi}^2[(1-w_{am})/\rho_L]^2$$

where the friction factor $(j_f)_{Li}$ is evaluated at the mean fractional vaporisation, w_{am} for the increment. Now w_{am} = 0.0233, therefore $(M_{Li})_{12}$ = 3.020 kg/s, Re_{Li} = 13465, hence $(j_f)_{Li}$ = 0.00445.

$$\therefore (\Delta P_f)_{12} = \frac{[4 \times 0.00445 \times 0.523 \times 7.8 \times 151.9^2 (1-0.0233)^2}{(0.02096 \times 709)} = 108 \text{ N/m}^2$$

Therefore pressure drop across first increment, $\Delta P_{12} = g(\rho_{TP})_{am} \Delta L_{12} + (\Delta P_f)_{12}$ = 1210 + 108 N/m² = 1318 N/m²

The total pressure drop across the remaining vaporisation increments were calculated as above and the results recorded in Table 4.13.

4.10.3.7 *Pressure balance over vaporiser*

As a further check that the desired circulation rate will be attained, a pressure balance will be completed to ensure that the effective static head provided by the thermo-syphon leg is sufficient to drive the circulating liquid against the frictional resistances in the equipment.

Effective static head = acceleration loss + total frictional resistance:

$$\rho_L g L_v - g \int_C^D \rho_{TP} \, dz = (M_T^2 \gamma_e / \rho_L S_E^2) + (\Delta P_f)_{AB} + (\Delta P_f)_{BC} + (\Delta P_f)_{CD} + (\Delta P_f)_{DA}$$

where the subscripts indicate the position in the vaporiser as shown in Figure 4.11.

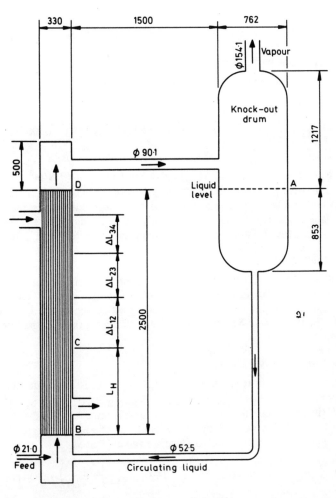

Figure 4.11 — *Arrangement and dimensions of vertical thermo-syphon vaporiser* (All dimensions mm).

Head provided by thermo-syphon leg, $\rho_L g L_v = 709 \times 9.81 \times 1.618$ N/m² = 11 253 N/m². Static head loss in vaporiser tubes,

$$g \int_C^D \rho_{TP}\, dz = g \sum (\rho_{TP})_{am} \Delta L.$$

From Table 4.13. $g \int_C^D \rho_{TP}\, dz = 2332$ N/m².

Effective Static Head = 11 253 − 2332 = 8921 N/m².

Acceleration Loss, $(M_T^2 \gamma_e / \rho_L S_E^2) = (3.092^2 \times 9.72)/[709 \times (6.379 \times 10^{-3})^2]$
= 3221 N/m²

Friction resistance in inlet pipe, $(\Delta P_f)_{AB}$ = 1370 N/m² (Section 4.10.3.4.)

Friction resistance in sensible heating section, $(\Delta P_f)_{BC}$ = 24 N/m² (Section 4.10.3.4)

Frictional resistance in vaporising section, $(\Delta P_f)_{CD}$ = 608 N/m² (Table 4.13).

Frictional resistance in exit pipe

$$(\Delta P_f)_{DA} = (M_T^2/\rho_L S_E^2)[\phi_e^2 4(j_f)_{LE}(L_E/D_E)(1-w_e)^2]$$
$$= (3.092^2)/[709 \times (6.379 \times 10^{-3})^2] 42.3 \times 4$$
$$\times 0.00243(1.5/0.09012)(1-0.14)^2$$
$$= 1677 \text{ N/m}^2.$$

Total Resistance = 3221 + 1370 + 24 + 608 + 1677 = 6899 N/m².

Since the driving pressure head exceeds the sum of the frictional resistances then the required circulation rate is assured and the design will be accepted taking into account the accuracy of the correlations used to describe the complex and interactive heat transfer and hydrodynamic behaviour of this type of vaporiser.

4.10.3.8 *Shell side heat transfer coefficient*

The heat transfer coefficient for the cooling of reaction product vapours on the shell side of the vaporiser was calculated using the method given in Section 4.10.4.3. A value of 180 W/m² K was obtained which approximates to the assumed value of 170 W/m² K in Section 4.10.3.1 and therefore the proposed design will be accepted.

4.10.3.9 *Total tube side resistance*

For evaluation of the heat transfer area, the total resistance on the tube side $[(x_w/k_w) + Rd_i + (1/hi)]_j^{-1}$ was assumed in Section 4.10.3.1. to be 1420 W/m² K. Although the inside coefficient varies with tube length, it is possible to weight the heat transfer coefficients according to the length of the increment for which they were calculated in order to obtain an average value. Hence,

$$h_i = h_{Li}\left(\frac{L_H}{L_v+L_H}\right) + (h_{vi})_{12}\left(\frac{L_{12}}{L_v+L_H}\right) + (h_{vi})_{23}\left(\frac{L_{23}}{L_v+L_H}\right) + (h_{vi})_{34}\left(\frac{L_{34}}{L_v+L_H}\right)$$

$$= [(750.7 \times 0.822) + (2528 \times 0.523) + (2948 \times 0.450) + (3509 \times 0.377)]/2.23$$

$$= 2057 \text{ W/m}^2\text{K}$$

Tube wall resistance $x_w/k_w = (0.0267 - 0.02096)/(2 \times 16.3)$ m² K/W
= 0.00018 m²K/W.

For alcohol vapours[20], $Rd_i = 0.0$, so

$$[(x_w/k_w) + Rd_i + (1/h_i)]^{-1} = [0.00018 + (1/2057)]^{-1} \text{ W/m}^2\text{K} = 1501 \text{ W/m}^2\text{K}$$

Comparison of this tube side coefficient with the assumed value of 1420 W/m² K shows that the latter was justified and no further iteration is required.

4.10.3.10 *Design of vapour-liquid separator*

Before the vapour enters the vapour superheaters, entrained liquid is to be removed using a knock-out drum. A simplified design procedure has been developed, employing nomographs, by Scheiman[25] and this will be applied here.

The first stage is to estimate the drum diameter

$$\begin{aligned}
\text{Liquid density, } \rho_L &= 44.15 \text{ lb/ft}^3 \ (\equiv 709 \text{ kg/m}^3). \\
\text{Vapour density, } \rho_v &= 0.15 \text{ lb/ft}^3 \ (\equiv 2.971 \text{ kg/m}^3). \\
\text{True vapour flow, } V &= 3436 \text{ lb/hr} \ (\equiv 0.433 \text{ kg/s}) \\
&= 3436/(0.15 \times 3600) \text{ ft}^3/\text{s} \\
&= 6.36 \text{ ft}^3/\text{s} \ (\equiv 0.00061 \text{ m}^3/\text{s}),
\end{aligned}$$

Also $R_d = (V/0.178D^2) [\rho_v/(\rho_L - \rho_v)]^{\frac{1}{2}}$ where D = drum diameter. R_d will be assumed equal to 0.3 as the recommended value for drums without mesh pad demisters is 0.5 maximum. Hence from nomograph[25] $D = 2.5$ ft ($\equiv 0.76$ m)

The total shell length is the sum of the vapour disengaging height required and the liquid holding length. The former part, for drums without mesh, is recommended by Scheiman to be 0.91 m for vapour space plus 0.305 m to the maximum liquid level. Therefore 1.22 m disengaging height will be used for the knock-out-drum.

The liquid holding length is based on the time required to empty the drum of liquid at the design flowrate. Therefore, it is necessary to specify the liquid holding time. Scheiman recommends between two and five minutes as a good working range and thus a holding time of two minutes will be used. Also for a fractional vaporisation of 0.14 the flowrate of hot 2-butanol is 19468 lb/hr. Therefore liquid flowrate = 19468/ (44,15 × 60) ft³/min = 7.35 ft³/min ($\equiv 0.825$ m³/s). Therefore, from second nomograph, liquid holding length = 2.8 ft ($\equiv 0.85$ m). So total shell length = 0.85 + 1.22 = 2.07 m.

Summarising, a knock-out drum of diameter 0.762 m and length 2.07 m will be attached to the vaporiser as shown in Figure 4.11

4.10.4 Design of vapour superheaters

Preliminary calculations showed that if one heat exchanger was used to superheat the alcohol vapour from 373 to 773 K, tubes longer than 5 m would be required introducing high costs for non-standard tubes and problems concerning structural supports. Therefore the vapour is to be heated using flue gas in two 1:1 exchangers with the vapours on the tube side to minimise pressure drop. Detailed design of the first superheater has been completed and the second only summarily designed since the procedure is identical in both cases.

4.10.4.1 First superheater

In this initial stage, the vapour is to be heated from 373 to 573 K using flue gas which enters at 673 K and cools to 423 K.

Considering a heat balance, $Q = mC_{pA}(t_1 - t_2)$, where $m = 1558.5$ kg/h, $C_{pA} = 2.038$ kJ/kgK at 473 K (Appendix A). Therefore $Q = 1558.5 \times 2.038 \, (573-373)$ kJ/h $= 0.6352 \times 10^6$ kJ/h.

But $Q = RC_p'(T_1 - T_2)$, therefore the flue gas flowrate, $R = (0.6352 \times 10^6)/[1.200 \times (673 - 423)]$ kg/h = 2116 kg/h.

Logarithmic mean temperature difference is given by,

$$\Delta T_{\ln} = \frac{(T_1 - t_1) - (T_2 - t_2)}{\ln[(T_1 - t_1)/(T_2 - t_2)]}$$

$$= \frac{(673 - 573) - (423 - 373)}{\ln[(673 - 573)/(423 - 373)]} \, K$$

$$= 72.1 \, K$$

Also assume design overall heat transfer coefficient, $U_D = 100$ W/m² K.

Then heat transfer area required is, $A = Q/U_D \Delta T_{\ln} = (0.6352 \times 10^6 \times 10^3)/(3600 \times 100 \times 72.1)$ m² = 24.47 m².

For a single tube pass exchanger containing 3/8 inch nominal size tubes ($d_o = 171.1$ mm; $d_i = 12.48$ mm) of 3.0 m length, surface area per tube $A_t = \frac{1}{2}\pi(d_o + d_i)L = \frac{1}{2}\pi(0.0171 + 0.01248) \, 3.0$ m² = 0.1394 m².

Thus number of tubes required = 176. Tube pitch, $P_T = 5d_o/4 = 0.02138$ m. Assuming 4 tie rods are required, $n_t = 176 + 4 = 180$, $n_d^2 \cong \frac{1}{3}(4n_t - 1) = 242$, so $n_d = 15.5$.

Shell diameter, $d_s = P_T(n_d + 1) = 0.02138 \, (15.5 + 1) = 0.354$ (13.9 inch), therefore selecting a standard shell diameter of 13 inches $d_s = 0.3302$.
Thus $n_d = d_s/P_T - 1 = (0.3302)/(0.02138) - 1 = 14.44$.

Now, $m = 0.577n_d + 0.423 = 8.76$, and $n_t = m(2n_d - m) = 8.76(14.44 \times 2 - 8.76) = 17$.

Nozzle allowance, $n_n = \frac{1}{2}(n_d + 1) = 8$. Therefore number of tubes, $N_t = n_t - n_n = 177 - 8 = 169$. So 165 tubes are available for heat transfer; 4 for tie rods.

Baffle spacing $B = 250$ mm.

The individual heat transfer coefficients will now be evaluated.

4.10.4.2 *Tube side coefficient*

Flow area $a_t = \frac{1}{4}(N_t \pi d_i^2)$

$= \frac{1}{4}[165 \times \pi \times (0.01248)^2]$ m^2

$= 0.0202$ m^2

Mass velocity, $G_t = m/a_t = 1558.5/(0.0202 \times 3600)$ kg/m^{2}s $= 21.45$ kg/m^{2}s.

Reynolds number, $Re_t = d_i G_t / \mu_A$

[where $\mu_A = 1.293 \times 10^{-5}$ kg/ms at 473 K (Appendix F)]

Thus $Re_t = (0.01248 \times 21.45)/(1.293 \times 10^{-5}) = 20703$.

Heat transfer[20] factor, $j_H = 70$

Thermal conductivity, $k_A = 0.040$ W/m K at 473 K (Appendix I)

Prandtl Number, $Pr = C_{pA}\mu_A/k_A$

Hence $Pr^{1/3} = [(2.038 \times 1.293 \times 10^{-5})/(0.040 \times 10^{-3})]^{1/3} = 0.870$.

Heat transfer coefficient, $h_i = j_H(k_A/d_i)(Pr)^{1/3}$
$= (70 \times 0.040 \times 0.870)/0.01248$ W/m^2 K $= 195$ W/m^2 K,

Correcting to the area associated with the centre of the tube wall, tube side coefficient $h_{iw} = 2h_i d_i/(d_o + d_i) = (2 \times 195 \times 0.01248)/(0.01248 + 0.0171)$ W/m^{2}K $= 164$ W/m^2 K.

4.10.4.3 *Shell side coefficient*

Tube clearance, $C'' = (P_T - d_o) = (0.02138 - 0.0171)$ m $= 0.00428$ m.

Flow area, $a_s = d_s C'B/P_T = (0.3302 \times 0.00428 \times 0.25)/0.02138$ m$^2 = 0.0165$ m^2.

Mass velocity, $G_s = R/a_s = 2116/(3600 \times 0.0165)$ kg/m^2 s $= 35.59$ kg/m^2 s.

Equivalent Diameter,

$$D_E = 8(0.43 P_T^2 - \tfrac{1}{8}\pi d_o^2)/\pi d_o \text{ m}$$

$$= 8(0.43 \times 0.02138^2 - \tfrac{1}{8}\pi \times 0.0171^2)/(\pi \times 0.0171) \text{ m}$$

$$= 0.0122 \text{ m}$$

Flue gas viscosity, $\mu_{fg} = 2.75 \times 10^{-5}$ kg/ms at 548 K.

Reynolds number, $Re_s = D_E G_s/\mu_{fg} = (0.0122 \times 35.59)/(2.75 \times 10^{-5}) = 15779$

Heat transfer[20] factor, $j_H = 70$

Thermal conductivity[20], $k_{fg} = 0.0458$ W/m K.

Prandtl number, $Pr = Cp_{fg}\mu_{fg}/k_{fg}$.

Hence $Pr^{1/3} = [(1.200 \times 2.75 \times 10^{-5})/(4.5845 \times 10^{-5})]^{1/3} = 0.896$.

Heat transfer coefficient, $h_o = j_H (k_{fg}/D_E) Pr^{1/3}$
$= 70 \times 0.0458 \times 0.896/0.0122$ W/m² K $= 235$ W/m² K.

Correcting to area associated with tube wall centre shell side coefficient h_{ow}
$= 2h_o d_o/(d_i + d_o) = 2 \times 235 \times 0.0171/(0.01248 + 0.0171)$ W/m² K $= 272$ W/m² K

Clean overall coefficient, $U_c = h_{iw} h_{ow}/(h_{iw} + h_{ow})$
$= 164 \times 272/(164 + 272)$ W/m² K $= 102.2$ W/m² K

Assume fouling factors $R_{di} = 0.0001$ for organic vapours $R_{do} = 0.0004$ for flue gas, tube wall resistance $x/k_w = 0.00231/16.29 = 0.000142$. Therefore overall design coefficient, $U_D = [1/U_c + x/k_w + R_{di} + R_{do}]^{-1}$
$= [1/102.2 + 0.000142 + 0.0001 + 0.0004]^{-1} = 96$ W/m² K.

This value is sufficiently close to the assumed value of 100 W/m² K to be acceptable.

4.10.4.4. Pressure drops

For the Tube Side, $Re_t = 20703$

Friction factor, $f = 0.03456$

Tube diameter, $d_i = 0.01248$ m

Number of passes, $n = 1$

Vapour specific gravity, $s = 0.001906$ (Appendix F)

Mass velocity, $G_t = 21.45$ kg/m² s

Tube length, $L = 3.0$ m

$\therefore$ Tube side pressure drop, $\Delta P_t = (fG_t^2 Ln)/1000 d_i s\phi_t$ N/m²
$= 0.03456 \times (21.45)^2 \times 3.0 \times 1/(1000 \times 0.1248 \times 0.001906 \times 1)$ N/m²
$= 2005$ N/m² $= 0.02$ bar.

For the Shell side, Mass velocity, $G_s = 35.59$ kg/m² s
Reynolds Number, $Re_s = 15779$
Friction factor, $f = 0.2808$
Specific gravity, $s = 0.000645$
Number of crosses, $(N+1) = 3000/250 = 12$
$\therefore$ Shell side pressure drop, $\Delta P_s = fG_s^2 d_s(N+1)/(1000 D_E s\phi_s)$
$= 0.2808 \times (35.59)^2 \times 0.330 \times 12/(1000 \times 0.0122 \times 0.000645 \times 1)$ N/m²
$= 179\,100$ N/m² $= 1.77$ bar.

4.10.5 Second superheater

This heat exchanger is required to heat the 2-butanol feed from 573 K to the reaction temperature of 773 K utilising flue gas which enters at 873 K and cools to 623 K.

Heat load, $Q = MC_{pA}(t_i-t_2)$
$= 1558.5 \times 2.674 \times (773-573)$ kJ/h $= 0.8335 \times 10^6$ kJ/h.

Flue gas requirement, $R = Q/C_{pfg}(T_1-T_2)$
$= 0.8335 \times 10^6/1.195 (873-623)$ kg/h $= 2790$ kg/h.

Logarithmic mean temperature difference, $\Delta T_{ln} = 72.1$ K, so heat transfer area required $= Q/U_D \Delta T_{ln}$ where overall design coefficent, $U_D = 100$ W/m² K. Therefore surface area required, $A = 0.8335 \times 10^6/(100 \times 10^{-3} \times 3600 \times 72.1)$ m² $= 32.1$ m².

Assuming dimensions of tubes and their layout identical to that selected for the first vapour superheater. Number of tubes required $= 32.1/0.1394 = 230$

Following the same procedure as described in Section 4.7.1, 223 tubes in 0.381 m diameter shell will achieve the necessary duty.

4.11 Utilisation of flue gas

Flue gas is used as the heating medium for the reactor and can then be usefully employed in the vapour preheaters which necessitates blending in order to achieve the desired temperatures and mass flowrates (Table 4.14)   Recycling of the flue gas is also desirable in the interests of energy economy.

Table 4.14 — *Summary of the flue gas requirements.*

Exchanger	Inlet Temperature (K)	Outlet Temperature (K)	Flowrate (kg/h)
Reactor	800	780	4124
Superheater 2	873	623	2790
Superheater 1	673	423	2116

If flue gas is available at 873 K, then the supply to the reactor will be provided by blending the latter with the exit gases from superheater number 2. The flowrates required were determined by a mass and heat balance assuming a datum temperature of 0°C and that the specific heat of flue gas remains constant.

Let the flowrate of flue gas at 873 K $= x$ and flowrate of flue gas at 623 K $= y$

Considering a mass balance, $x + y = 4124$; heat balance, $873x + 623y = (800 \times 4124)$. Hence $x = 2919$ kg/h, $y = 1205$ kg/h.

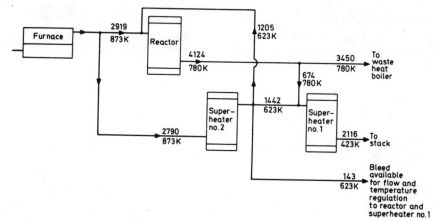

Figure 4.12 — *Utilisation of flue gas.*

By completing similar mass and heat balances, a network of flue gas utilisation is proposed which is illustrated in Figure 4.12. The exit flue gas from superheater number 1 is suitable for direct discharge to the atmosphere and the remainder of the reactor exit gases is to be passed to a waste heat boiler for steam raising. Also the small excess of exit flue gas from superheater number 2 will be available for flow and temperature control of the inlet streams to the reactor and superheater number 1.

Chapter 5

THE CONDENSER

5.1 General discussion

The products of reaction are to be discharged from the reactor at 370°C and passed to the shell side of the thermo-syphon reboiler type evaporator to evaporate the feed. Details of the design of this unit are presented in section 4.10.3 where it was shown that the majority of the heat in the reaction products has been utilised and that the vapour temperature of the reaction products has been reduced to 125°C; a suitable temperature to enter the condenser. This temperature is considerably above the dew point of the vapours so that no condensation will occur in the reboiler. Hence the composition of the vapours entering the condenser will be that given in Table 5.1 which were obtained from the output columns of Table 4.1

Table 5.1 – *Composition of vapours entering condenser.*

Stream	Weight (kg/h)	Weight Fraction	kg Moles per hour	Mole Fraction
MEK	1351.30	0.8670	18.768	0.4737
Hydrogen	37.50	0.0241	18.600	0.4684
Alcohol	169.74	0.1089	2.294	0.0579
Total	1558.54	1.0000	39.663	1.0000

Inspection of Table 5.1 shows that nearly 50% of the vapours entering the condenser are noncondensable. Consequently the heat transfer rate will be low and will vary along the length of the condenser tubes. Hence the design procedure developed by Colburn and Hougen [26] will be utilised for this condenser.

5.2 Material balance over condenser

The vapours enter the condenser at 125°C and at a pressure of 1.0 bar. They will be cooled until more than 80% of the condensables have been removed from the vapour. The actual amount will depend on the exit temperature of the vapours leaving the condenser which, in turn depends on the temperature and the rate of the cooling water entering the condenser. However it is not possible to make a complete material balance over this condenser and then design the unit to comply with the stated specifications because the Colburn-Hougen procedure necessitates a series of step-wise calculations. Consequently the composition of the condensate and the effluent vapour will not be known until the design analysis has been completed.

5.3 Preliminary specifications

It is proposed to instal a horizontal single pass shell side, single pass tube side heat exchanger operating under counter-current flow conditions, and from some approximate preliminary calculations it would appear that a suitable unit would be one about 2.5 m in length containing 104 tubes, each 0.0191 m o.d., 16 s.w.g. in thickness arranged on a 0.0254 m square pitch. Baffles will be introduced with a baffle spacing of 0.35 m.

5.4 Estimation of heat transfer area

5.4.1 *Water rate*

McAdams [27] stated that for most heat exchange operations the economic cooling water rate was of the order of 1.0 m/s and usually this rate was sufficient to minimise the deposition of silt. Therefore the condenser design will be based on a water velocity through the tubes of 1.0 m/s. On this basis the quantity of water required can be estimated as follows:

Flow area per tube = 1.948×10^{-4} m^2. Then water rate = $1.948 \times 10^{-4} \times 104 \times 1.0 \times 3600 \times 1000 = 72932$ kg/h of cooling water.

5.4.2 *Temperature rise of the water*

At this stage in the design of the condenser it is not possible to make an overall heat balance in order to calculate the exit water temperature because the amount of vapour condensed will not be known. Neither will the exit temperature of the gas phase be known. Therefore it will be necessary to make a preliminary estimate of the temperature rise of the cooling water, which subsequently will be verified.

Generally water for industrial processes is available at 24°C for most duties; the maximum outlet temperature from most heat exchangers is restricted to 50°C, since at higher temperatures there is a tendency for air bubbles to be liberated from the water which may initiate corrosion. For condensation duties of the kind encountered here it is essential that the temperature difference be maintained as great as possible throughout the unit because the condensation rate will be limited by the presence of the non-condensable gas-hydrogen. Furthermore it is essential that the temperature of the outside tube surface — that is the condensate film — is below the dewpoint temperature of the vapours. Since it is envisaged that this condenser will operate under counterflow conditions the limiting exit water temperature is determined by the dew point temperature of the entering vapour. This is 62.5°C.

Hence for safety the condensate film temperature at the entrance section of the condenser will be taken to be 56°C. Then, since the gas side heat transfer coefficient will be of the order of 100 W/m^{2}K and the MEK partial pressure difference between the gas phase at 125°C and condensate surface at 56°C will be about 0.15 bar, the heat transfer rate to the water in this section will be of the order of 30000 W/m^2. The water side coefficient will be about 1000 W/m^2 K and therefore the exit water temperature will be approximately $30000/1000 = 56 - t = 26$°C. These estimates are very approximate. However a temperature rise of 2°C will be accepted initially and confirmed later.

5.4.3 Tube-side heat transfer coefficient

The resistance to heat transfer is the sum of the resistances of the non-condensable gas, condensate film, pipe wall, scale deposit and cooling water; and since the mass velocity of the vapour changes from point to point throughout this unit the gas film resistance will vary accordingly. However, all the other resistances will remain fairly constant with the result that they may be grouped together to form a composite tube side coefficient. Furthermore in order to reduce the number of calculations in the design of this item the "tube side coefficient" will be evaluated at the mean cooling water temperature of 25°C. Each of the film coefficients contributing to this composite coefficient will now be evaluated.

5.4.3.1 Condensate film coefficient

The film coefficient of heat transfer for the condensation of MEK and alcohol on the outside of a single horizontal tube may be obtained from the Nusselt Equation[20]:

$$h_c = 0.72 \left[\frac{k^3 \rho^2 g \lambda}{D \mu \Delta t_f} \right]^{1/4} \qquad (5.1)$$

where

k = Thermal Conductivity of Condensate = 0.1456 W/m K
ρ = Condensate density = 775 kg/m³
μ = Condensate viscosity = 4.53 kg/m s
λ = Latent heat of Condensate = 455041 J/kg
Δt_f = Temperature difference = 14°C

Then, taking the chosen tube dimensions of the condenser

$$h_c = 0.72 \left[\frac{(0.1456)^3 (775.0)^2 (455041)(9.81)}{(4.53 \times 10^{-4})(0.0191)(14.0)} \right]^{0.25}$$

$$= 2059 \text{ W/m}^2 \text{ K}$$

Kern[20] states that in a horizontal condenser, the splashing of condensate as it drops over successive rows of tubes tends to increase the heat transfer coefficient. Therefore the above estimate should be checked for the proposed condenser of length 2.5 m containing 104 tubes. This may be done by using the chart developed by Kern[20] where

$$G'' = W/LN^{2/3} = 1521/(2.5 \times 104^{0.67}) = 27.51 \text{ kg/h m} \qquad (5.2)$$

Then, using the physical properties quoted above

$$h_c = 3123 \text{ W/m}^2 \text{ K}$$

5.4.3.2. Water film coefficient

The cooling water film coefficient will be estimated at the mean water temperature. The physical properties of water used in the following calculation are those of water at 25°C. Then the water Reynolds number = $(0.0157 \times 1.0 \times 1000)/(0.81 \times 10^{-3}) = 19383$.

Then from Kern (Figure 24), $j_H = 67$ and the Prandtl Number for the water film = $Cp\mu/k = (4.183 \times 10^3 \times 0.81 \times 10^{-3})/0.6159 = 5.05$, and $Pr^{0.33} = 1.72$.

Then

$$h_T = (k/d_1)Pr^{0.33} j_H = (0.6159 \times 1.72 \times 67)/0.0157$$
$$= 4521 \text{ W/m}^2 \text{ K based on inside tube diameter}$$

This value compares favourably with the value of 4545 W/m² K obtained graphically from Figure 25 of Kern and will therefore be accepted.

5.4.3.3. Tube wall heat transfer coefficient

$$h_w = \frac{2k}{d_o \ln(d_o/d_i)} = \frac{2.0 \times 45.0}{0.0191 \ln(0.0191/0.0157)} = 24037 \text{ W/m}^2 \text{ K}$$

where 45.0 W/m K is the Thermal Conductivity of Steel.

5.4.3.4. Scale resistance

The total scale resistance has been estimated to be $R_d = 3.523 \times 10^{-4} \text{ m}^2\text{K/W}$. Then the total tube side heat transfer coefficient is

$$\frac{1}{h_{cw}} = \frac{1}{h_c} + \frac{d_1}{d_o h_T} + \frac{1}{h_w} + R_d \qquad (5.3)$$

$$= \frac{1}{3123} + \frac{0.0157}{0.0191 \times 4521} + \frac{1}{24037} + 3.523 \times 10^{-4}$$

$$= 8.96 \times 10^{-4} \text{ m}^2/\text{K W}$$

or $h_{Tn} = 1116 \text{ W/m}^2 \text{ K}$

5.4.4. The shell-side coefficient

The mass velocity of the vapour will change considerably from point to point as it passes through the shell side of the condenser and therefore it is necessary to estimate the incremental heat transfer area as the condensation progresses using the Colburn-Hougen stepwise procedure.

Starting from the hot gas entrance:

Gas Entry Temperature		=	125°C
At the entrance, heat transferred	q_1	=	0
Shell-side Flow Area	a_s	=	$D_i \times C' \times B/P_T$ (Kern p138)
where Internal Diameter of Shell	D_i	=	0.3366 m
Clearance between tubes,	C'	=	0.0063 m
Baffle Spacing	B	=	0.3500 m
Tube Pitch	P_T	=	0.0254 m

Then

$$a_s = \frac{0.3366 \times 0.0063 \times 0.35}{0.0254} = 0.0292 \text{ m}^2$$

The Shell Equivalent Diameter is

$$D_E = 4\left(\frac{P^2 - \frac{1}{4}\pi d_0^2}{\pi d_0}\right) = 4\left(\frac{0.0254^2 - 0.7854 \times 0.0191^2}{3.1416 \times 0.0191}\right) = 0.0239 \text{ m}$$

where d_o is the external diameter of the condenser tubes.

Now a_s and D_E are constant throughout the condenser, therefore $G = 1558.54/3600 \times 0.0292 = 14.82 \text{ kg/m}^2 \text{ s}$, and $Re = GD_E/\mu = (14.82 \times 0.0239)/(9.59 \times 10^{-6}) = 36934$ where the viscosity of the gaseous mixture was obtained from Appendix F.
The Prandtl Number for the gas mixture was obtained as shown in Table 5.2.

Table 5.2 — Mean heat capacity of mixture.

Component	Weight fraction	Heat capacity (J/kg K)	Heat Content
MEK	0.8670	1664.00	1442.69
Hydrogen	0.0241	14650.00	350.07
Alcohol	0.1089	1760.00	191.67
Total	1.0000		1984.43

That is the heat capacity of the gas phase at the entrance to the condenser is 1984.43 J/kg K.
Then the Prandtl Number is $Pr = Cp\mu/k = 1984.43 \times 9.59 \times 10^{-6}/0.0604 = 0.315$. This value of the Prandtl Number is very low for gases and therefore must be confirmed. This has been done using the correlation proposed by Euchen's and recommended by Gambill [19]

$$Pr = \frac{C_p(M)}{C_p(M) + 1.0383 \times 10^4} \tag{5.4}$$

where $C_p(M)$ is the molar heat capacity in kg moles. That is

$$Pr = \frac{1.4308 \times 10^5}{1.4308 \times 10^5 + 1.0383 \times 10^4} = 0.93$$

This value is high but in the range of Prandtl Numbers normally expected for gases. Furthermore it is only slightly less than the Prandtl Number of the reaction mixture at 45% conversion (see paragraphs 4.7.3.). This suggests that one of the prediction procedures presented in the appendices is unreliable when applied to this mixture when it contains a high concentration of hydrogen. In fact, the calculation of the Prandtl Number by equation (5.4) suggests that the heat capacity estimation is reasonable. It was based on Dobratz' method[19] and checked by the Group Contribution method, and the agreement is seen to be satisfactory. The heat capacity estimate will therefore be accepted. However Friend & Adler [28] state that the procedure for predicting the viscosities of mixtures is unreliable when the system contains appreciable quantities of hydrogen. Furthermore, since the prediction of the thermal conductivity makes use of the predicted viscosity, the estimation of both of these physical properties, for this system, must be considered further before being accepted for the design calculations.

The procedure for estimating viscosity, presented in Appendix F, was used to calculate the viscosity of the mixture in the first increment of the condenser analysis at 394 K, and the results are summarised in Table 5.3, from which $\mu_m = 9.59 \times 10^{-6}$ kg/m s.

Table 5.3 — Viscosity of mixture entering condenser.

Component	Mole Fraction	Molecular weight	$yM^{1/2}$	$\mu y M^{1/2} \times 10^5$
Hydrogen	0.4690	2.016	0.6651	0.7985
MEK	0.4732	72.100	4.0223	3.7407
Butyl alcohol	0.0578	72.120	0.4985	0.4337
Total	1.0000		5.1859	4.9729

In addition the weighted mean viscosity was calculated for this mixture and, on this basis, predicted viscosity was found to be 9.3×10^{-6} kg/m s. Hence there is very little difference in the predicted values by the alternative calculation procedures.

The same series of calculations was performed for the evaluation of the thermal conductivity by the two methods described above and the results are summarised in the Tables 5.4. and 5.5.

Table 5.4 – Thermal conductivity of mixture entering condenser.

Component	Thermal Conductivity	Mole Fraction	$yM^{1/2}$	$kyM^{1/2}$
Hydrogen	0.2206	0.4690	0.5917	0.1305
MEK	0.0173	0.4732	1.9713	0.0341
Butyl alcohol	0.0183	0.0598	0.2432	0.0045
Total		1.0000		0.1691

Table 5.5 – Weight mean thermal conductivity.

Component	Weight Fraction	Thermal conductivity	ky
Hydrogen	0.0240	0.2206	0.0053
MEK	0.8670	0.0173	0.0150
Butyl alcohol	0.1089	0.0183	0.0020
Total	1.0000		0.0203

From Table 5.4 k_m = 0.0604 W/mK. In this case the weighted mean thermal conductivity is from Table 5.5: k_m = 0.0203 W/mK. Hence the difference is considerable and insertion of the weighted mean thermal conductivity in the Prandtl Number gives

$$Pr = \frac{C_p\mu}{k} = \frac{1984.43 \times 9.59 \times 10^{-6}}{0.0203} = 0.94$$

The agreement between the Prandtl Number evaluation using the predicted physical properties, including a weighted mean thermal conductivity and Euchen's correlation is very good. Furthermore the Prandtl Number for hydrogen itself is 0.9 and therefore the above value of the Prandtl Number will be used in subsequent calculations. Also in future calculations involving thermal conductivity of the vapours of this system the weighted mean will be calculated. Then inserting these values into Kern (Figure 28) gives

$$h_0 = \frac{120 \times (0.94)^{0.33} \times 0.0203}{0.0239} = 100 \text{ W/m}^2\text{ K}$$

The shell side heat transfer coefficient at point 1 (the gas entrance).

The rate of mass transfer of the MEK and the 2-butanol through the gas film can be obtained from the heat transfer coefficient by applying the Colburn[26] j_m- Factor concept. Thus:

$$k_g = \frac{h_0}{C_p M_m P_{g\,\text{ln}}} \left(\frac{Pr}{Sc}\right)^{0.667} \qquad (5.5)$$

where for point 1:

M_m = the mean molecular weight of the gas = 39.06
Sc = The Schmidt Number = $(\mu/\rho D)$
$P_{g\,\text{ln}}$ = Log mean partial pressure of the noncondensable gas

The Diffusivity of the condensables can be estimated from Gilliland's [29] Equation:

$$D = \frac{1.0 \times 10^{-7} \times T^{1.75}}{\Pi[V_A^{1/3} + V_B^{1/3}]^2} \left[\frac{1}{M_A} + \frac{1}{M_B}\right]^{0.5} \qquad (5.6)$$

where

M_A is the molecular weight of the diffusing component.
M_B is the molecular weight of the non-diffusing component.
Π is the total gas pressure
V_A is the molecular volume of the diffusing component.
V_B is the molecular volume of the non-diffusing component.

In the present analysis both MEK and the alcohol diffuse through the permanent gas hydrogen. To reduce the amount of calculation a composite diffusivity of the condensable mixture will be evaluated from equation (5.5.). Thus, the mean molecular weight of the condensables at the gas entrance is 72.3 and the atomic volumes quoted by Perry [19] give the following molecular volumes:-

$$V_{MEK} = (4 \times 16.5) + (8 \times 1.98) + 5.48 = 87.3 \text{ ml/g mole.}$$
$$V_{ALC} = (4 \times 16.5) + (10 \times 1.98) + 5.48 = 91.28 \text{ ml/g mole.}$$
$$V_{H_2} = 7.07 \text{ ml/g mole} = V_B$$

Then the molecular volume of the condensables is $V_A =$ 87.7 ml/g mole. At the entrance to the condenser, and after the condensate film has been formed the gas film temperature will not be very different from that of the bulk of the gas. Therefore with a gas entry temperature of 125°C i.e. 398 K, the physical properties will be estimated at this temperature. Then

$$D = \frac{(394)^{1.75} \times 10^{-7}}{1.0 \times [87.7^{0.33} + 7.07^{0.33}]^2} \left[\frac{1}{72.3} + \frac{1}{2.016}\right]^{0.5}$$
$$= 0.614 \times 10^{-4} \text{ m}^2/\text{s}$$

The density of the gas phase at the above temperature is: $\rho = 12.564\, M_m/T$ where M_m is the mean molecular weight of the gas, i.e. $M_m = 39.06$, and $\rho = 12.564 \times 39.06/394 = 1.246$ kg/m^3.

The Schmidt Number, $Sc = \mu/\rho D = (9.59 \times 10^{-6})/(1.246 \times 0.614 \times 10^{-4}) = 0.1254$. This value of the Schmidt Number will be assumed to be constant throughout the condenser.

Then inserting the above values into equation (5.5) gives

$$k_g = \frac{100.0}{1984.43 \times 39.06} \left(\frac{0.94}{0.1254}\right)^{0.667} \frac{1}{P_{g\,\text{ln}}} = \frac{0.0049}{P_{g\,\text{ln}}}\,\text{s/m}$$

Finally the heat balance proposed by Colburn & Hougen[26] can be expressed

$$h_o(T_g - T_f) + k_g M_m \lambda(\rho_v - \rho_f) = h_{ow}(T_f - T_w) = U\Delta T$$

where

- h_o = Shell side vapour coefficient, W/m^2 K
- h_{ow} = Tube side coefficient based on outside diameter W/m^2 K
- p_f = Partial pressure of condensables in condensate film, bar.
- T_f = Condensate Interface temperature, K
- T_g = Gas temperature, K.
- T_w = Water temperature, K
- λ = Latent heat of condensables, J/kg

POINT 1

At this point $T_g = 398$ K, $T_w = 299$ K, $p_v = 53864$ N/m^2, $\lambda = 455296$ J/kg. Let $T_f = 327$ K, then the vapour pressures at T_f K are

$$P_{\text{MEK}} = \frac{101325}{760} \times 10^{\left(6.97421 - \frac{1209.6}{T_f - 57}\right)} \tag{5.7}$$

$$P_{\text{ALC}} = \frac{101325}{760} \times 10^{\left(7.21 - \frac{1158.34}{T_f - 104.94}\right)} \tag{5.8}$$

The partial pressure of the condensables in the condensate film p_f will be obtained from the values for the pure components using Raoult's Law. Thus from Table 5.1 the mole fraction of MEK in the vapour in the entering gas on a hydrogen-free basis is 0.88 and in order to start the calculation the condensate composition will be taken to be 0.88 mole fraction MEK and 0.12 mole fraction alcohol. This estimate will be checked and if necessary revised.

Then at $T_f = 327$ K from equations (5.7) and (5.8) $P_{MEK} = 41601$ N/m² and $P_{ALC} = 13139$ N/m², so that:

$$p_f = (0.88 \times 41601) + (0.12 \times 13139) = 38184 \text{ N/m}^2$$

$$p_{gf} = 101325 - 38184 = 63141 \text{ N/m}^2$$

$$p_{gv} = 101325 \times 0.4684 = 47460 \text{ N/m}^2$$

$$p_{g\,ln} = \frac{63141 - 47460}{\ln(63141/47460)}$$

$$k_g = 0.0049/54928 = 8.9 \times 10^{-8}$$

Substituting these values in the heat balance gives $100\,(398-327) + 8.9 \times 10^{-8} \times 39.06 \times 455296\,(53864-38184) = 7100 + 24874 = 31974$ W/m² $= 1116(327-299) = 31248$ W/m²

This difference corresponds to an error of 0.6 K and therefore the trial will be accepted. Hence at point 1: $U\Delta T_1 = 0.5\,(31974 + 31248) = 31611$ W/m². At point 1, $q = 0$.

POINT 2

Let the gas have cooled to 385 K at this point. Then assuming that condensation occurs at constant pressure, reduction in the partial pressure of the condensables will be proportional to the moles of each component condensed between point 1 and point 2. Then the number of moles of MEK condensed will be $97885 \times 18.7680/356368 = 5.16$ kg moles.

Table 5.6 — *Gas condensation rate in first increment.*

Component	Vapour Pressure 398K	Vapour Pressure 385K	ΔP	Mass Present	(kg Moles) Condensed	Remaining
MEK	356368	258483	97885	18.768	5.16	981.14
2 Butanol	241175	158612	82563	2.294	0.79	112.00
Hydrogen				18.564	–	37.50
Total				39.626	5.95	1130.64

The number of moles of 2-butanol condensed in the interval was estimated in a similar manner and the results are summarised in Table 5.6. The composition of the condensate in the first increment will be for MEK $(5.16/5.95) = 0.87$ mole fraction MEK. This compares with the mole fraction of 0.88 assumed above. The difference is so small that the assumed liquid composition will be accepted. Then the heat losses in the interval are:

1. Heat loss in Condensing MEK   = (370.16 × 443140) J/h
                                     = 1.6403 × 10^8 J/h
2. Heat lost in Condensing alcohol = (42.99 × 562150) J/h
                                     = 3.2459 × 10^7 J/h
3. Heat lost in Cooling hydrogen   = (37.5 × 14392 × 13.0) J/h
                                     = 7.0161 × 10^6 J/h
4. Heat lost in Cooling MEK   = (98144 × 1650 × 13.0) J/h
                                     = 2.1045 × 10^7 J/h
5. Heat lost in Cooling alcohol   = (112.0 × 1745 × 13.0) J/h
                                     = 2.5407 × 10^6 J/h

   Total Heat Loss   = 2.2706 × 10^8 J/h

Drop in Temperature of the water during interval = (2.2706 × 10^8)/(72932 × 4183.3) = 0.74 K,

Then

   New Water Temperature   = (299 − 0.74) = 298.3 K
   New Gas Rate (from Table 5.6)   = 1130.64 kg/h
   Viscosity of mixture   = 9.74 × 10^{-6} kg/m s
   Heat capacity of mixture   = 2028 J/kg K
   Thermal Conductivity of mixture
   (based on weighted mean)   = 0.0232 W/m K

Then the

   Reynolds Number = (0.0239 × 1130.64)/(9.74 × 10^{-6} × 0.0292 × 3600)
                   = 26392

   Prandtl Number = (2028 × 9.74 × 10^{-6})/0.0232 = 0.85

and $Pr^{0.33}$ = 0.95 and $Pr^{0.67}$ = 0.8973.

Then from Kern (Figure 28)

$$h_o = (98 \times 0.95 \times 0.0232)/0.0239 = 90.37 \text{ W/m}^2 \text{ K}$$

Applying equation (5.4) with the following data: T_g = 385 K; M_A = 74.59 (from Table 5.6), M_B = 2.016; V_A = 87.69; V_B = 7.07, gives the Diffusivity: D = 5.9 × 10^{-5} m^2/s and the density: ρ = 1.15 kg/m^3.

Therefore the Schmidt Number Sc = (9.74 × 10^{-6})/(1.15 × 5.9 × 10^{-5}) = 0.1436 and $Sc^{0.67}$ = 0.2742.

The Mass Transfer Coefficient is:

k_g = (90.37 × 0.8973)/(2028 × 35.32 × 0.2742 × p_{gln}) = 0.0041/p_{gln}

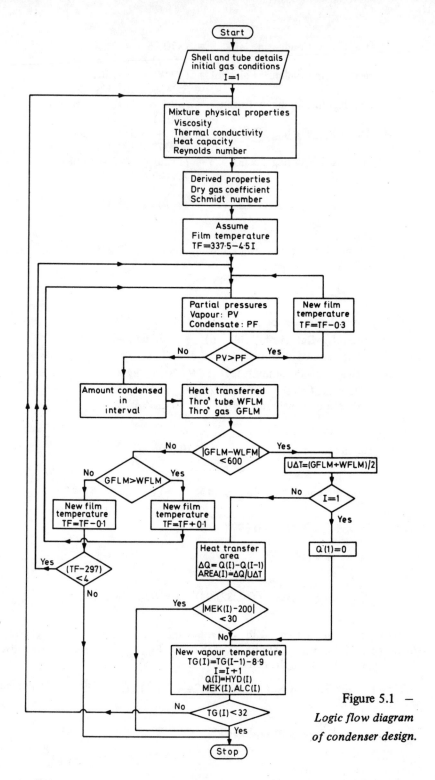

Figure 5.1 — *Logic flow diagram of condenser design.*

Table 5.7 – Summary of stepwise analysis of condenser.

Vapour Temperature (K)	Water Temperature (K)	Film Temperature (K)	Vapour loads Alcohol (kg)	Vapour loads MEK (kg)	Vapour loads Hydrogen (kg)	Total (kg)	Gas film Coefficient (W/m²K)	Mass transfer Coefficient (m/s) × 10²	$\frac{1}{U\Delta T}$ × 10⁴	$\frac{1}{U\Delta T_m}$ × 10⁴	Increment Heat load (W) × 10⁴	Area (m²)
398.0	299.0	327.0	169.70	1351.3	37.5	1558.5	100.0	8.9	0.316	0	–	
385	298.3	322.0	112.00	981.14	37.5	1130.64	90.37	6.8	0.389	0.353	6.3072	2.260
380.2	298.0	315.8	94.20	861.8	37.5	993.5	84.0	5.6	0.505	0.447	3.693	1.653
371.3	297.7	310.7	68.2	675.2	37.5	780.8	77.4	4.75	0.694	0.599	2.959	1.775
362.4	297.4	306.5	48.2	521.5	37.5	607.2	72.7	4.09	0.998	0.846	2.347	1.986
353.5	297.2	303.3	33.3	396.6	37.5	467.4	69.4	3.57	1.484	1.241	1.843	2.288
344.7	297.1	301.1	22.3	296.6	37.5	356.4	67.1	3.14	2.261	1.873	1.434	2.685
335.7	297.0	299.6	14.5	217.7	37.5	269.7	65.7	2.76	3.464	2.862	1.106	3.165

Then

$$90.37(385 - T_f) + \frac{0.0041}{p_{g\,ln}} \times 35.32 \times 455296(48002 - p_f)$$

$$= 1116(T_f - 300.45) = U\Delta T_2 \quad (5.9)$$

In order to solve equation (5.9) it is necessary to assume a condensate composition in the interval and then, if required, amend subsequently. Thus let the mole fraction of MEK in the condensate be 0.85. Let $T_f = 322$ K then $p_f = 30601$ N/m², $p_{gf} = 70724$ N/m², $p_{gv} = (0.5248 \times 101325) = 53171$ N/m², $p_{g\,ln} = 61614$ N/m²,

Then:

$$90.37(385 - 322) + \frac{0.0041 \times 35.32 \times 455296}{61614}(48002 - 30601) = 5693 + 18557$$

$$= \mathbf{24250} \simeq 1116(322 - 300.45) = \mathbf{24029} \text{ W/m}^2$$

The agreement between each side is good and a temperature of 322 K will be accepted. Then at point 2, $U\Delta T_2 = (24250 + 24049)/2.0 = 24149$ W/m². The Heat load is: 6.3072×10^4 W and the surface area in the interval is: $6.3072 \times 10^4/27879 = 2.266$ m².

The stepwise procedure described above was continued by computer and the logic diagram for the program is presented in Figure 5.1. The results obtained are summarised in Table 5.7 which shows that a total heat transfer area of 15.812 m² is required. This area was obtained by iterating until an inlet cooling water temperature of 24°C was attained when the computer calculations were terminated.

In Section 5.3 it was decided to instal a single pass tube side, single pass shell side condenser containing 104 tubes, each 0.0191 m, o.d. Since the above heat transfer calculations were based on the outside tube diameter, the surface area per metre length of condenser is: 6.22 m²/m. Hence, the length of condenser required for this duty is: 15.812/6.22 = 2.55 m. Therefore, for safety, a condenser containing 104 tubes 0.0191 m, o.d., 2.55 metres long will be recommended. The detailed specifications of the unit are presented in Table 5.8. There it will be seen that 0.375 m has been allowed for the inlet and outlet water headers so that the overall length of the condenser is 3.30 m.

Table 5.8 – *Condenser Specification*

Overall Length	3.30 m	Inlet Header	0.373 m
Shell Diameter	0.438 m	Outlet Header	0.373 m
Tube Length	2.55 m	Passes Tube Side	1.0
Number of Tubes (Square pitch)	104	Passes Shell Side	1.0
		Number of Baffles	8
Tube i.d.	0.0158 m	Baffle Spacing	0.35 m
Tube o.d.	0.0191 m		

5.5. Pressure drop through condenser
5.5.1 *Tube side pressure drop*
The tube side pressure drop ΔP_T may be estimated from:

$$\Delta P_T = \frac{fG^2 nL}{2\rho d_i \phi_T} \qquad (5.10)$$

where ΔP_T = Pressure drop, N/m²
  f = Friction factor, Dimensionless
  G = Flow rate, kg/m² s
  L = Tube length = 2.55 m
  d_i = Tube diameter = 0.0158 m
  ρ = Water = 1000 kg/m³
  ϕ_T = Viscosity ratio = 1.0

Then $G = 72932/(3600 \times 0.0203) = 997.97$ kg/m² s and the Reynolds Number = 19383 (see Section 5.4.3.). Then from Kern (Figure 27) $f = 0.0331$, and $\Delta P_T = 0.0331 \times 997.97^2 \times 1.0 \times 2.55/2 \times 1000 \times 0.0158 \times 1.0 = 2660.$ N/m².

5.5.2 *Shell side pressure drop*
The shell side pressure drop will be based on the mean flow rate of the gas through the shell using the equation

$$\Delta P_S = \frac{fGd_s(N+1)}{2\rho D_E \phi_S} \qquad (5.8)$$

where
  N = Number of Baffles = Tube length/Baffle spacing = 2.55/0.35 = 8.
  $\bar{\phi}_S$ = Viscosity Correction = 1.0
  D_E = Shell Equivalent Diameter = 0.2039 m

Then
	Inlet	Outlet	Mean
Reynold's Number	36934.	6322.	
Friction Factor	0.23	0.33	0.28
Gas Flow Rate (kg/m² s)	14.82	2.57	8.70
Gas Density (kg/m³)	1.246	0.39	0.818

Then
$$\Delta P_S = \frac{0.28 \times 8.7^2 \times 0.3366 \times 8.0}{2 \times 0.818 \times 0.0239 \times 1.0} = 1460 \text{ N/m}^2$$

The pressure drop through the tubes on the shell and tube side are quite normal for the duty envisaged and therefore the design of this is acceptable.
Finally, the overall mass balance on the condenser is summarised in Table 5.9

Table 5.9 — *Overall material balance on condenser.*

Component	Input vapour (kg/h)	(%)	Output liquid (kg/h)	(%)	Output vapour (kg/h)	(%)
MEK	1351.30	86.70	1133.60	87.96	217.70	80.72
2- Butanol	169.74	10.89	155.24	12.04	14.50	5.38
Hydrogen	37.50	2.41	–		37.50	13.90
Total	1558.54	100.00	1288.84	100.00	269.70	100.00

The liquid discharged from the condenser at 26.6°C, say 27°C, will be pumped to an intermediate storage to be mixed with the distillate from the solvent recovery still of the extraction unit. This mixture will be the feed to the product distillation unit. The mass balance over this storage unit is presented in Section 7.7.9 and the design of the distillation plant is discussed in Chapter 8. The vapour discharged from the condenser at 62.7°C, say 63°C, is to be passed to the absorption unit. The analysis of this unit is discussed in Chapter 6. This concludes the design of the condenser.

Chapter 6

THE ABSORPTION COLUMN

6.1 General discussion

The gas phase discharged from the cooler-condenser at 62.7°C containing 0.1384 mole fraction of MEK and 0.009 mole fraction of 2-butanol in hydrogen is to be treated in an absorption column with a very dilute aqueous solution of MEK recycled from the extraction column. The unit is to be designed to absorb 98% by weight of the MEK and nearly all the alcohol, and must take the following criteria into consideration.

(i) The heat effects associated with the dissolution of MEK and alcohol in the absorbent liquor.

(ii) The system MEK/water/1.1.2 trichlorethane forms an isopicnic at 11.0 % by weight of MEK.

The heat effects are considered below. The feed to the extraction column will be the effluent liquor discharged from the absorption column and in order to avoid the complications referred to in Section 3.1 the concentration of MEK in the liquor discharged from the absorption column must be about 10% by weight. On this basis the material balance for the absorption column is presented in Table 6.1. In this table it will be seen that the liquor fed to this column contains 0.50% by weight of MEK. This consists of 1932.8 kg/h of 0.52% weight MEK solution discharged, as raffinate, from the extraction column and 10.35 kg/h of make-up water to replace that leaving with hydrogen discharged from the absorber. This quantity of vapour may be estimated as follows.

The absorbent liquor enters the absorber at 27°C and comes into contact with the exit gas. Since this gas will have passed through the absorption column it will be assumed that it is saturated with water vapour at the temperature of the entering liquid. Then at 27°C the vapour pressure of water is 3.565 kN/m² and the molecular weight of the exit gas is 2.276 so that the humidity of the exit vapour will be

$$\mathcal{H} = \frac{3.565}{101.325 - 3.565} \left(\frac{18.0}{2.276}\right) = 0.2875$$

$$= 2.88 \text{ kg per kg dry gas}$$

Water vapour discharged with gas from the absorber = 42.5 × 0.2875 = 12.25 kg.

Then the total gas rate entering the absorber will be (from Table 6.1) 21.8162 kg moles per hour at 335.7 K corresponding to a volumetric flowrate of 600.92 m³/h. Therefore the absorption column required for this duty will be small. However, it will be seen from Table 6.1 that 3.146 kg moles of MEK and alcohol are to be

Table 6.1 – *Material balance on absorption column.*

	Input				Output			
Component	Vapour Feed		Irrigating liquor		Gas phase		Effluent liquor	
	(kg/h)	(% wt)	(kg/h)	(% wt)	(kg/h)	(% wt)	(kg/h)	(% wt)
MEK	217.70	80.72	9.65	0.50	4.50	8.22	222.85	10.32
2-butanol	14.50	5.38	–	–	0.50	0.92	14.00	0.65
hydrogen	37.50	13.90	–	–	37.50	68.49	–	–
water	–	–	1935.05	99.50	12.25	22.37	1922.80	89.03
Total	269.70	100.00	1944.70	100.00	54.75	100.00	2159.65	100.00

absorbed with the result that a considerable amount of heat will be released to the water. Hence it is necessary to assess whether the absorbent liquor must be cooled during the absorption process as it passes through the column in order to maintain a satisfactory absorption rate. Inter-stage cooling is most conveniently achieved in a plate column but because of the low throughput envisaged, a packed column is preferred here. The MEK and 2-butanol absorbed in the water release latent heat and heat of solution to the irrigating liquor while the water evaporated in humidifying the gas will take heat out of the system. These heat exchangers will affect the absorption rate and therefore the design of the unit.

6.2 Analysis of heat exchanges in absorption column

If no inter-stage cooling is provided the temperature of the liquor leaving the base of the column will be

(i) Heat of Condensation of MEK: $(213.20 \times 443.14) = 94477.45$ kJ/h

(ii) Heat of Condensation of 2-butanol: $(14.0 \times 560.0) = 7840.0$ kJ/h

(iii) Heat of solution: (227.20×0.35) $= 79.52$ kJ/h

(iv) Heat in cooling gases from 62.7°C to 27°C:
$[(4.5 \times 1.4701) + (0.5 \times 1.53) + (37.5 \times 14.65)] \times (62.7 - 27)$
$= 19876.17$ kJ/h

Total heat released $= 122273.14$ kJ/h

(v) Heat removed by water vapour: $(12.25 \times 2437.90) = 29864.28$ kJ/h

(vi) Heat gained by irrigating liquor:
(a) Water: $1922.80 \times 4.186 \times (t - 27)$ $= 8050.38 (t - 27)$

(b) MEK: $222.85 \times 2.299 \times (t - 27)$ $= 512.33 (t - 27)$

(c)   Alcohol: $14.0 \times 2.429 \times (t-27)$ $= 34.01 (t-27)$

Total heat gained by liquor $= 8596.72 (t-27)$ kJ/h

Then neglecting heat losses through the walls of the column, the temperature leaving the absorber will be: $122273.14 = 29864.28 + 8596.72 (t-27)$ or $t = 37.8°C$

This temperature is not excessive and since the design, construction and operation of this unit will be less complicated if the effluent liquor is cooled after it leaves the absorber and before entering the extraction unit it is proposed that no inter-stage cooling be introduced.

The variation in temperature was estimated throughout the absorption column by calculating the temperature rise for each 10.0% of MEK absorbed in the liquid. In order to make the calculations tractable the following assumptions were made. (1) The gas phase was rapidly saturated with water vapour on entering the column, and the heat exchanges associated with this transference of water vapour were assumed to occur during the first 10.0% absorbed from the gas at the base of the column. (2) The vigorous contact between gas and liquid throughout the column results in the gas and liquid having the same temperature at any plane in the column. In practice there will be a small difference but the heat exchanges associated with this difference will be negligible.

The incremental calculations at each 10% MEK absorbed were performed in the same manner as those presented above in this section for the entire column. Therefore the details will not be repeated here. The results obtained are summarised in Table 6.2 where it will be seen that the temperature increases by about 1.3°C for each 10.0% MEK absorbed in the liquor. Table 6.2 also presents the partial pressure of MEK in the bulk of the gas phase and at the interface. The interfacial partial pressures were estimated from the data of Scheibel and Othmer[30] which has been reproduced in Figure 6.1. Inspection of the last three columns shows that the driving force will be positive throughout the column, although it becomes very small at the top of the column. This however is not a temperature effect, but is due to the low concentration of MEK in the gas phase leaving the absorber. Table 6.2 may be used to estimate the height of packing required for the absorption duty specified in Table 6.1. This height depends on the interfacial area which in turn is related to the choice of packing. Consequently packing selection will now be considered.

6.3 Estimation of column diameter

The diameter of this absorption column depends on the choice of packing, the gas and liquid flowrates and the height of packing required for the proposed duty. The absorption duty is not very demanding but the liquid flowrate is low because of the concentration constraint. Therefore the height of packing will probably be large, necessitating a strong packing element and for this reason a ceramic packing will be recommended. Furthermore the gas flowrate has been calculated to be

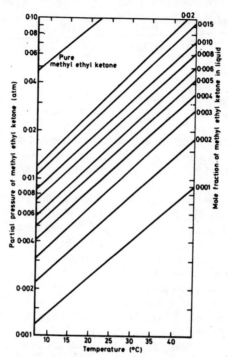

Figure 6.1 — *Equilibrium partial pressure of methyl ethyl ketone over aqueous solution.*
[Based on Scheibel & Othmer[30] by permission of AIChE]

600.92 m³/h so that in all probability the column diameter will be of the order of 0.5 m. Hence, since the system to be processed is non-corrosive, random packed stoneware Raschig rings will be a suitable packing[31]. Also on the basis of the anticipated column diameter, rings of size 3.8 cm would be appropriate to prevent channeling. Then by Morris & Jackson[31] the economic gas rate will be in the range:-

$$\text{Minimum: } 1850(1.2946/0.5520)^{0.33} = 2451 \text{ m}^3/\text{h m}^2$$

$$\text{Maximum: } 2900(1.2946/0.5520)^{0.33} = 3842 \text{ m}^3/\text{h m}^2$$

where $\rho_{AIR} = 1.2946$ kg/m³ at NTP and $\rho_{gas} = 0.552$ kg/m³ at NTP at the entrance. Taking the mean of these rates the column cross sectional area will be:—

$$600.92/3146.5 = 0.191 \text{ m}^2 \equiv 0.493 \text{ m diameter}$$

Therefore a diameter of 0.5 m will be selected. Then the volumetric liquid rate at the base of the column will be:

$$2159.65/979.4 = 2.205 \text{ m}^3/\text{h}$$

where 979.4 kg/m³ is the density of the effluent liquor. Similarly at the top of the column the liquid rate will be

$$1944.7/1000.0 = 1.945 \text{ m}^3/\text{h}$$

where 1000 kg/m³ is the density of the water entering the absorber.
The superficial wetting rate is then

At the bottom	At the top
$\dfrac{2.205}{0.196} = 11.25$ m/h:	$\dfrac{1.945}{0.196} = 9.924$ m/h

and the wetting rate is

At the bottom	At the top
$\dfrac{11.25}{125.0} = 0.09$ m³/m h:	$\dfrac{9.924}{125.0} = 0.08$ m³/m h

where 125.0 m²/m³ is the surface area per unit volume of the packing. That is, the wetting rate for this packing is always above the minimum of 0.08 m²/m³ so that the packing will be completely wetted throughout the column. Therefore the proposed absorption column diameter of 0.5 metre will be accepted.

6.4 Gas and liquid column loading

In order to ensure that this absorption column operates within specification it is necessary to confirm the loading state of the column. This may be done by Leva's chart, Figure 6.2 [19] and since the flowrate of gas and liquid is greatest in the lower sections of the column, Leva's correlation will be applied to the base of the absorber. For this column

Fractional voidage[31] ε, = 0.73

$$G = \frac{269.70}{0.196 \times 3600} = 0.38 \text{ kg/m}^2 \text{ s}$$

$$L = \frac{2159.65}{0.196 \times 3600} = 3.06 \text{ kg/m}^2 \text{ s}$$

Liquid viscosity in centipoise μ, = 0.98 cP

$\rho_L = 979.0$ kg/m³

$$\rho_g = \frac{269.70}{600.92} = 0.45 \text{ kg/m}^3$$

$$\psi = \left[\frac{\text{Density of Water}}{\text{Density of Irrigating Liquid}}\right] = \frac{1000}{979} = 1.02$$

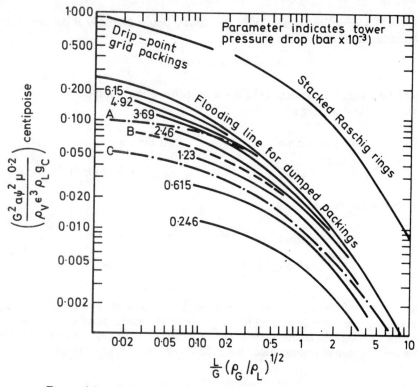

Figure 6.2 – *Generalised flooding and pressure drop correlation.*
[Based on Leva[19] by permission of AIChE]
  A, approximate upper limit of loading zone.
  B, line representing majority of data.
  C, approximate lower limit of loading zone.

Then Leva's Parameter = $[G^2 a\psi^2 \mu^{0.2}/\rho_g \varepsilon^3 \rho_L g]$ which has the dimensions of $(ML^{-1}T^{-1})^{0.2}$ and therefore the viscosity is best expressed in centipose. That is

$$\frac{0.38^2 \times 125.0 \times 1.02^2 \times 0.98^{0.2}}{0.45 \times 0.73^3 \times 979.0 \times 9.807} = 0.011$$

and the abscissa of Leva's chart is

$$\frac{L}{G}\left(\frac{\rho_g}{\rho_L}\right)^{0.5} = \frac{3.06}{0.38}\left(\frac{0.45}{979.0}\right)^{0.5} = 0.1726$$

The value of this parameter at loading for an ordinate of 0.011 is 0.85 from Figure 6.2. Therefore the fractional loading = 0.172/0.85 = 20.2 %. That is under normal steady-state conditions with the proposed flow rates the column will operate at 20.0% loading.

As will be seen later, this low loading is desirable during "start-up operations". This is discussed in Chapter 12.

6.5 Height of packing

The height of packing required for the duty proposed depends on the overall mass transfer coefficient, the interfacial area and the driving force. The interfacial area per unit volume of absorption column has been established from the choice of the packing. Thus, in section 6.3 it was reported to be 125.0 m²/m³. Since a column of diameter 0.5 m has been confirmed the interfacial area per unit height of column will be $0.196 \times 125.0 = 24.5$ m² per metre height of packing.

The mean driving force may be evaluated from Table 6.2 using Simpson's Rule thus

$$\Delta p_m = \tfrac{1}{30} [(0.0008+0.0574)+4(0.0066+0.0183+0.0268+0.0338+0.0512) \\ +2(0.0111+0.0260+0.0368+0.046)] = 0.028 \text{ bar}.$$

6.5.1 *The mass transfer coefficient*

The values of the mass transfer coefficient depends on the resistance to mass transfer in the gas and liquid phases. Morris and Jackson claimed that this depended on the ratio (ρ_s/HP) where

ρ_s is the density of the soluble gas at the process temperature

H is the solubility coefficient of the solute gas, kg/m³ bar

P is the total pressure, bar.

When the value for this index is less than 5.0×10^{-4} the transport process is gas phase controlled. When the index is greater than 0.2 it is liquid phase controlled and between these two limits both phases contribute. For the system MEK/water at the mean temperature of the absorber

$$\rho_s = 2.737 \text{ kg/m}^3$$
$$H = 859.0 \text{ kg/m}^3 \text{ bar from Figure 6.1}$$
$$P = 1.0 \text{ bar}$$

and

$$\rho_s/HP = 2.737/(859 \times 1.0) = 3.2 \times 10^{-3}$$

which suggests that both phases contribute to the transport process. This agrees with the findings of Othmer and Scheibel [30] who studied the absorption of MEK from air into water in a laboratory column packed with glass Raschig rings. Their results showed that the liquid phase offered considerable resistance to mass transfer. However, it is questionable whether their results can be scaled-up to the duty and equipment considered here. For this design it is proposed to apply the well known, well established, correlations for the gas and liquid film mass transfer coefficients and to use Othmer and Scheibel's correlation for comparison.

Table 6.2 – *Temperature variation and driving force in absorption column.*

Gas flowrate		Fraction absorbed	Liquor flowrate		Temperature	Partial pressure		ΔP (bar)
(kg/h)	Mole Fraction MEK		(kg/h)	Mole Fraction MEK	(K)	Gas phase (bar)	Interface (bar)	
63.20	0.0036	1.0	1944.70	0.0013	300.0	0.0036	0.0028	0.0008
84.52	0.0186	0.90	1953.77	0.0040	301.3	0.0186	0.0120	0.0066
105.84	0.0331	0.80	1975.09	0.0068	302.6	0.0331	0.0220	0.0111
127.16	0.0473	0.70	1996.41	0.0095	303.9	0.0473	0.0290	0.0183
148.48	0.0610	0.60	2017.73	0.0123	305.1	0.0610	0.0350	0.0260
170.80	0.0748	0.50	2040.05	0.0149	306.4	0.0748	0.048	0.0268
193.12	0.0888	0.40	2062.37	0.0177	307.7	0.0888	0.0520	0.0368
215.94	0.1028	0.30	2085.19	0.0205	310.3	0.1028	0.0690	0.0338
239.76	0.1170	0.20	2109.01	0.0233	310.5	0.1170	0.0710	0.0460
264.58	0.1312	0.10	2133.83	0.0260	310.7	0.1312	0.0800	0.0512
269.70	0.1474	0	2159.65	0.0298	310.8	0.1474	0.0900	0.0574

Many correlations for the film coefficients of mass transfer have been developed. These are generally based on the evaluation of the constants and exponents of a dimensional analysis and of these, the correlations proposed by Morris and Jackson have been applied extensively and are well accepted. Consequently, they will be utilised for these design calculations, and the value of the gas and liquid film coefficient will be determined at the mean column conditions corresponding to 50% of the solute gases absorbed. From Table 6.2, this corresponds to the point where the temperature in the column is 306.0 K and all the parameters required for estimation of the mass transfer coefficients will be evaluated at this temperature. The flowrates and partial pressures will also be abstracted from the data in Table 6.2 corresponding to the gas containing 0.0748 mole fraction of MEK.

The gas and liquid film coefficients will now be calculated for these conditions.

6.5.2 *Gas phase mass transfer coefficient*

The gas phase coefficient of mass transfer is conveniently correlated by the equation [31]

$$k_g = 36.1 R_g C \rho_{rs} V^{0.75} \left[\frac{P}{(P-p)_m} \right] \left(\frac{1}{P} \right)^{0.25} \left(\frac{293}{T_f} \right)^{0.56} \quad (6.1)$$

where:

- R_g Packing factor for the gas film = 2.70
- ρ_{rs} Density of MEK vapour at 293 K = 3.00 kg/m³
- C Gas mixture constant

$$C = (\rho_r/\mu_r)^{0.25} D_r^{0.5} \quad (6.2)$$

- ρ_r Density of gas mixture = 3.16×10^{-4} gm/ml at the plane where 50% of the MEK has been absorbed.
- μ_r viscosity of gas mixture at the same plane = 8.9×10^{-6} poise
- D_r Diffusivity of MEK vapour in hydrogen at 293 K = 0.56 cm²/s.

$$C = \left(\frac{3.16 \times 10^{-4}}{8.9 \times 10^{-5}} \right)^{0.25} \times 0.56^{0.5} = 1.03$$

- V Gas velocity relative to liquid surface

$$V = \frac{V_g}{3600\varepsilon} + V_c \quad (6.3)$$

- V_g Gas flowrate = 524.5 m³/h at midpoint plane
- ε Voidage = 0.73 for the selected packing
- V_c Liquid flowrate = 0.05 m/s from Figure 15B of Morris and Jackson[31]

Then

$$V = \left(\frac{524.5}{0.196 \times 3600 \times 0.73}\right) + 0.05 = 1.07 \text{ m/s}$$

$$T_f = 306.0 \text{ K}$$

$\dfrac{P}{(P-p)_m}$ The drift factor.

In this factor P is the total pressure and $(P-p)_m$ is the log mean partial pressure of the hydrogen in the bulk of the gas and at the interface. For the conditions in Table 6.2 $(P-p)_m = 0.9594$ and the Drift Factor = 1.042

Then

$$k_g = 36.1 \times 2.70 \times 1.03 \times 3.00 \times (1.07)^{0.75} \times 1.042 \times (293/306)^{0.56}$$
$$= 322.3 \text{ kg/h m bar}$$

Applying Othmer and Scheibel's correlation (equation 15), i.e. $k_g a = 0.307 D_G G^{0.8}$ where the symbols denote British units gives a value of $k_g = 410$ kg/h m² bar.

The agreement between Othmer & Scheibel's predication and the general correlation is quite good. However, since in this design the gas film coefficient is required for the transfer of MEK through hydrogen whereas the Othmer correlation is based on the diffusion of MEK through air it would be prudent to take the mean value of these two predications. Then

$$k_g = 366 \text{ kg/h m}^2 \text{ bar}$$

6.5.3 Liquid phase mass transfer coefficient

The liquid phase mass transfer correlation developed by Morris and Jackson for liquid film coefficients in packed columns requires the evaluation of a number of constants and coefficients which would have to be "guestimated" from a very speculative basis for the MEK/water system. Hence it will be more realistic to apply the correlation by Sherwood and Holloway[33] and the form presented by Norman[32] appertains directly to 3.8 cm stoneware Raschig rings:

$$k_L a/D = 228.0 (L/\mu)^{1-0.22} (\mu/\rho D)^{0.5} \qquad (6.4)$$

where

D Diffusivity of MEK in the liquid = 0.33×10^{-5} m²/h

L Liquid flowrate = $2040.05/0.196 = 10408.42$ kg/m² h

μ Viscosity 2.88 kg/m h

ρ 990 kg/m³ is the density of the liquor at 50% absorption at 33°C

Then

$$\frac{k_L \times 125.0}{0.33 \times 10^{-5}} = 228.0 \left(\frac{10408.42}{2.88}\right)^{0.78} \left(\frac{2.88}{990 \times 0.33 \times 10^{-5}}\right)^{0.5}$$

$$k_L = 228.0 \times 595.98 \times 29.69 \times 0.33 \times 10^{-5}/125.0$$

$$= 0.1065 \text{ m/h}$$

The value predicted by the Othmer-Scheibel correlation specifically for the system MEK/water, namely $k_L a = 599 D_L L^{0.8}$ (where the terms should be expressed in British Units) gives a value of $k_L = 0.078$ m/h.

The agreement is very good but again it is considered prudent to take the mean value. That is, the value of k_L to be used in this design will be $k_L = 0.092$ m/h.

6.5.4 The overall mass transfer coefficient

The overall mass transfer coefficient is estimated from the relation

$$1/K_G = 1/k_g + 1/H k_L \qquad (6.5)$$

where H Slope of p.v.c. equilibrium curve, = 859 kg/m³ bar from Figure 6.1

and

$$1/K_G = 1/366 + 1/(859 \times 0.092) = 0.0027 + 0.0127 = 0.0154$$

$$K_G = 65.1 \text{ kg/h m}^2 \text{ bar}$$

6.6. Absorption column height

The height of packing required to absorb 217.7 kg/h of MEK and the associated 2-butanol may be estimated from the equation: $217.70 = k_G a \cdot \Delta p_m \, l$ where a; the interfacial area per metre height of packing = 24.5 m². From section 6.5, $\Delta p_m = 0.028$ bar. That is the recommended height of packing for the specified duty will be

$$l = 217.7/(65.1 \times 24.5 \times 0.028) = 4.87 \text{ metres} \simeq 5.0 \text{ metres}$$

The height of packing specified is large for the duty envisaged. This is due to the low driving force resulting from the rise in temperature in the irrigating liquid as it passes through the column and to the low loading and consequently the low liquid film coefficient. This height of packing will be accepted nevertheless and the reserve capacity will be advantageous during start-up; as discussed in Chapter 13.

6.7 Pressure drop over absorption column

The pressure drop over the packing can be estimated from Leva's chart using the values of the coordinates in section 6.4. From this chart the pressure drop per metre of packing is 8.17 N/m^2 per metre of packing. Then the total pressure drop = $(8.17 \times 24.0)/10^5$ = 0.0019 bar. That is the pressure drop over the packing is very small.

6.8 Specification

The above analysis has shown that the absorption column required for the duty stated in Table 6.1 will be a packed column 0.5 m diameter, 5.0 m packed height. The packing recommended for this unit will be stoneware rashig rings 3.8 cm diameter, 3.80 cm high with a wall thickness of 0.5 cm. Under normal operation this unit will be expected to operate at about 20% loading.

The effluent gas discharged from this absorber at the rate of 54.75 kg/h will be fed to the reactor for regeneration of the catalyst. However it is recommended that a flame trap be provided in the line but the detailed specification of this equipment is outside the scope of this design study.

The liquor discharged from the base of this absorption column will leave at a temperature of 37.8°C (say 38°C). At this temperature the vapour pressure of the MEK exceeds 0.25 bar and must be reduced before entering the solvent extraction unit. Consequently it is proposed to insert a water cooled heat exchanger between the absorption column and extraction unit. The design of this unit is straight forward and has been performed in the same way as in the heat exchanger calculations presented in section 4.7. The specifications of this heat exchanger are:

1. Heat load required to reduce liquor temperature to 27°C: 94964 kJ/h
2. Absorber liquor to be cooled by water and water temperature rise restricted to 4°C. That is exit water temperature 28°C.
3. Water flowrate 5600 kg/h
4. Suitable heat exchanger would be one 0.305 m diameter containing 109 tubes each 1.905 cm o.d.
5. Estimated heat transfer coefficient 900 W/m^2 K.
6. Estimated heat transfer area 16.23 m^2
7. Estimated length of heat exchanger 2.5 m.

This concludes the design of the absorption unit.

Chapter 7

THE SOLVENT EXTRACTION UNIT

7.1 General discussion

The effluent liquor discharged from the gas absorption column at 2159.65 kg/h will be treated with 1.1.2 trichlorethane to extract the MEK and 2-butanol. The extract phase containing the MEK and alcohol will be distilled to recover the products and the solvent will be recycled to the extraction column. All the MEK cannot be

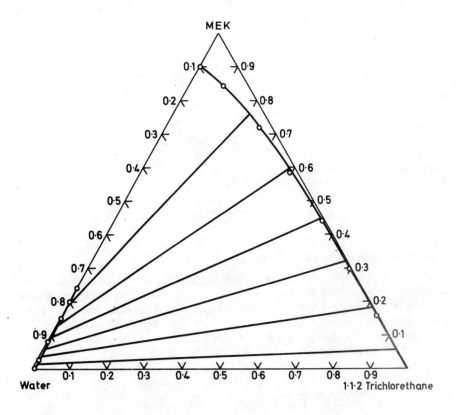

Figure 7.1 — *Phase equilibrium diagram for the system methyl ethyl ketone/ water/ 1.1.2 trichlorethane.*

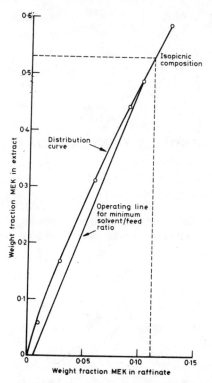

Figure 7.2 — *Distribution diagram.*

extracted from the aqueous solution, but examination of the phase equilibrium diagram, Figure 7.1 and particularly the distribution diagram, Figure 7.2 shows that it is possible to recover sufficient MEK to leave a final raffinate containing only 0.5% weight MEK. This will be the basis for the material balance over this unit.

7.2 Solvent feed ratio and material balance

In order to evaluate a material balance over this unit the solvent/feed ratio must be estimated. Thus, a feed solution of 10.3% weight of MEK in water and a final raffinate of 0.5% weight form the basis for the design of the extraction unit. These limits have been inserted in Figure 7.2 where the approximate operating line corresponding to the minimum solvent to feed ratio has also been drawn. This line intersects the distribution curve at $x = 0.1032$. Then the

$$\frac{\text{Solvent}}{\text{Feed}} \simeq \frac{E}{R} = \frac{x_f - x_a}{y_1 - y_{n+1}} = \frac{0.1032 - 0.005}{0.48 - 0.0} = 0.205$$

and the minimum solvent rate = $2159.65 \times 0.205 = 443$ kg/h.

The optimum solvent/feed ratio may be estimated approximately from the procedure developed by Treybal[6] which will be followed here.

The volumetric flowrate of the feed is

$$222.85 \text{ kg of MEK} \equiv \frac{222.85}{806.0} = 0.2765 \text{ m}^3$$

$$14.00 \text{ kg of 2-butanol} \equiv \frac{14.0}{807.0} = 0.0173 \text{ m}^3$$

$$1922.8 \text{ kg water} \equiv \frac{1922.80}{1000.00} = 1.9228 \text{ m}^3$$

Feed volumetric flowrate $= 2.2166 \text{ m}^3/\text{h}$

The minimum volumetric solvent flowrate is $0.3343 \text{ m}^3/\text{h}$ which is only 15.0% of the feed volumetric flowrate and therefore a solvent flowrate of twice the minimum will be used as a first estimate, *i.e.* $880 \text{ kg/h} = 0.664 \text{ m}^3/\text{h}$. Then $R_1 = 0.664/2.2166 = 0.30$.

The uninstalled cost of the extraction column can be estimated approximately from the relation

$$C_V = \frac{5000 \, a^{0.7}}{\text{stage or compartment*}} \tag{7.1}$$

*could be unit of a mechanically agitated column

where a is the sectional area of the column in m^2.

Let the diameter of the column be 20.0 cm. This will be confirmed later (see section 7.3.2). Then $a = 0.7854 \times 0.2^2 = 0.0314 \text{ m}^2$ and $C_V = 5000 \times 0.0314^{0.7} = 443.6$ Since the calculations are approximate let $C_V = £445.0$. Then

$$C_E = C_V(p/Y + b) \tag{7.2}$$

where

C_E is Annual cost of an extraction stage including equipment and operating costs.

p is (cost of instruments, piping and installation as a fraction of extractor costs) + 1.0 *i.e.* $p = 2.5$ (see Treybal[6]).

Y is Payout time = 5 years

b is Annual maintenance costs as a function of equipment costs = 0.25 (from Treybal).

$$C_E = 445.0(2.5/5.0 + 0.25) = £334.0 \tag{7.3}$$

Let E_{OE}:— Fractional overall stage efficiency for extraction be 0.90.

Now
$$B = O(1+r)$$

where

O Rate of distillate flow in extract recovery column = 213.2/72.1 = 2.957 kg/moles h

r, reflux ratio in extract recovery column = 2.5 (see section 7.7.).

Then
$$B = 2.957(1+2.5) = 10.35 \text{ kg moles/h.}$$

As a check

$$B = U\left(1+\frac{\beta}{\alpha-1}\right) + \frac{R_1 q \rho_s}{M_s(\alpha-1)} \qquad (7.4)$$

where

U Rate of extraction in kg moles of MEK per hour = 2.957 kg moles/h
α Relative voliatility between MEK and solvent = 3.5 approximately from vapour pressures over extractor concentration.
$\beta = r/r_{min}$ for extract recovery column = 2.0 (see section 7.7.)
q Volume rate of flow of raffinate = 2.216 m³/h
ρ_s Density of solvent = 1325.0 kg/m³
M_s Molecular weight of solvent kg/kg mole = 133.4

Then

$$B = 2.957\left(1.0 + \frac{2.0}{3.5}\right) + \frac{0.17 \times 2.216 \times 1325.0}{133.4(3.5-1)}$$

$$= 4.65 + 1.50 = 6.15 \text{ kg moles/h}$$

Taking the mean
$$B = 8.25 \text{ kg moles/h}$$

Evaluating

$$J = 1.0 + \left[\frac{4.0 C_E}{(C_F - C_S/m) F q H E_0 (mR-1) \ln(mR)}\right] \qquad (7.5)$$

where

C_F concentration of MEK in feed = 100.54 kg/m³
C_s concentration of MEK in recycled solvent: $c_s/m \to 0.0$
F value of extractable solute = £0.2545 per kg
H time of operation = 8000 h per annum
m distribution coefficient = 5.7 (mean ratio).

Then

$$J = 1.0 + \left[\frac{4.0 \times 334.0}{(100.54 - 0.0)0.2545 \times 2.216 \times 8000 \times 0.9(1.71 - 1.0) \ln 1.71} \right]$$

$$= 0.0086 + 1.0 = 1.0086$$

Now number of ideal extraction stages

$$n_E = \frac{\text{Log } 1.0 - 2.0/(1 \pm \sqrt{J})}{\text{Log } (mR)} - 1.0 = 9.0 \tag{7.6}$$

Now

$$A = \left(\frac{C_D \gamma n_{D\,min}}{E_D G_D} + \frac{C_h}{G_h} \right) \left(\frac{p}{Y} + b \right) + C_{hc} H \tag{7.7}$$

where

C_D uninstalled cost of a distillation tray and accompanying shell = £2000
γ $= n_D/n_{D\,min}$
n_D number of ideal distillation stages for extract recovery = 20 (see section 7.7).
$n_{D\,min}$ minimum n_D = 6.0 (see section 7.7)
G_D allowable velocity of MEK in Still = 73.0 kg moles/h m²
C_h uninstalled cost of heat transfer equipment = £33.0 per metre ³.
G_h vapour handling capacity of heat transfer equipment = 0.5 kg moles/h
C_{hc} cost of steam and coolant = £0.08 per kg mole

$$A = \left(\frac{2000 \times 20.0 \times 6.0}{0.7 \times 73.0 \times 6.0} + \frac{33.0}{0.5} \right) \left(\frac{2.5}{5.0} + 0.25 \right) + 0.08 \times 7000$$

$$= (782.8) \times 0.75 + 560.0 = £1196.5 \text{ kg mole}$$

Then for an extraction column

$$X = \frac{Aq\rho_s\varepsilon}{M_s(\alpha-1)} + qlSH \tag{7.8}$$

where

ε = 1.0 since MEK is more volatile than the solvent.
l = fraction of solvent lost per stage = 0.0005
S = value of solvent = £350/m³

$$X = \frac{1196.5 \times 2.2166 \times 1325.0 \times 1.0}{133.4 \times (3.5-1.0)} + 2.2166 \times 0.0005 \times 350 \times 8000$$

$$= 10537.1 + 2715.3 = £13252.4$$

and the value of the abscissa in Treybal's chart is

$$\frac{XE_{OE}(R_1+1)^g}{n_E R_1 C_E} = \frac{13252.4 \times 0.9(0.17+1.0)^{0.45}}{9.0 \times 0.30 \times 334.0} = 9.83$$

where g has been taken to 0.45. Treybal states $g \to 0.45$.
Then from Figure 7.3 based on Treybal[6] the optimum solvent feed ratio = 0.28 and the optimum Solvent Rate = 0.28 × 2.216 × 1325.0 = 823 kg/h. This value is close to the initial estimate and will be accepted.

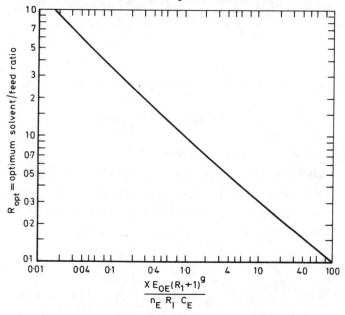

Figure 7.3 — *Correlation of optimum solvent/feed ratio.*
[Based on Treybal[6] by permission of the publishers, McGraw-Hill Book Co.]

The optimum solvent rate has been estimated from very approximate cost data obtained from different sources and where necessary indexed. Mumford and Jeffreys have shown [34] that variation in solvent rate with costs about the minimum is small. That is, there is not a sharp optimum so that a solvent rate of 900 kg/h will be recommended. This solvent will be regenerated from the extract by distillation and recycled to the extraction column. It will contain small amounts of water and MEK, and from a material balance over the solvent recovery unit the solvent flowrate will be 900.77 kg/h. Since this solvent will leave the bottom of the solvent recovery still it must be cooled and a cooler will be installed to deliver the regenerated solvent to the extractor at 27°C. That is the solvent extraction column will be designed to operate isothermally at 27°C. However since the phase equilibrium data is available at 25°C this will not be corrected. Then, on the basis of a solvent flowrate of 900.77 kg/h and the MEK content of the final raffinate restricted to 0.5% MEK, the composition of the final extract is obtained graphically from Figure 7.1 thus: 79.6% trichlorethane: 19.00% MEK: 1.24% alcohol: 0.14% water. On this basis the flowrate of extract and raffinate leaving the column is: $E_N + R_N = 3060.42$ and $0.005 R_N + 0.1900 E_1 = 224.05$ or:

$$E_1 = 1128.11 \text{ kg/h and } R_N = 1934.30 \text{ kg/h}$$

Table 7.1 — *Material balance over extraction column.*

Component	Input				Output			
	Feed		Solvent		Raffinate		Extract	
	(kg/h)	(% wt)	(kg/h)	(% wt)	(kg/h)	(% wt)	(kg/h)	(% wt)
MEK	222.85	10.32	1.20	0.14	9.65	0.50	214.40	19.00
2-butanol	14.00	0.65					14.00	1.24
Water	1922.80	89.03	1.57	0.17	1922.80		1.57	0.14
1.1.2 TCE	—		898.00	99.67	1.86		896.14	79.62
Total	2159.65	100.00	900.77	100.00	1934.30	100.00	1126.11	100.00

These flowrates are the basis of the material balance on the extraction column which is presented in Table 7.1 This balance will be the basis for the following design calculations.

7.3 Evaluation of dimensions of extraction column

The principal dimensions of the extraction column will be estimated from the equation

$$213.2 = K_G a (C^* - C)_m h \qquad (7.9)$$

where

K_G is the overall mass transfer coefficient
a is the interfacial area per metre height of column
h is the height of the column
$(C^* - C)_m$ is the mean overall driving force.

Each of these terms will be evaluated. The mean overall driving force may be obtained from Figure 7.2 while the other terms in equation (7.9) depend on the type of column selected for the required duty. There are many different types of extraction column employed for chemical processing and it is not possible to make a comparison in this report. Here it is proposed that a rotating disc contactor (RDC) be installed because the extraction duty is not large and a small unit will suffice. In addition the RDC is very versatile and will operate efficiently under conditions slightly different from the design specifications. Each of the terms in equation 7.9 will now be estimated.

7.3.1 *Estimation of the mean driving force.*

The system MEK/water/1.1.2 trichlorethane is a type-2 partially miscible system and, although reflux cannot be applied to obtain a concentrated extract because of the existence of the isopicnic, it is necessary to obtain the coordinates of the operating line graphically because of its curvature. This has been done by drawing straight lines from the difference point in Figure 7.5 to intersect the equilibrium curve on the raffinate and extract side of the diagram. The coordinates located in this way are plotted on the distribution diagram Figure 7.2 and the mean driving force for the selected process conditions was estimated from this diagram by applying Simpson's rule. Thus from Figure 7.4 the mean driving force

$$\Delta y = \frac{1}{12}[(0.030+0.301)+4(0.110+0.247)+2(0.185)]$$

$$= \frac{2.129}{12} = 0.1774$$

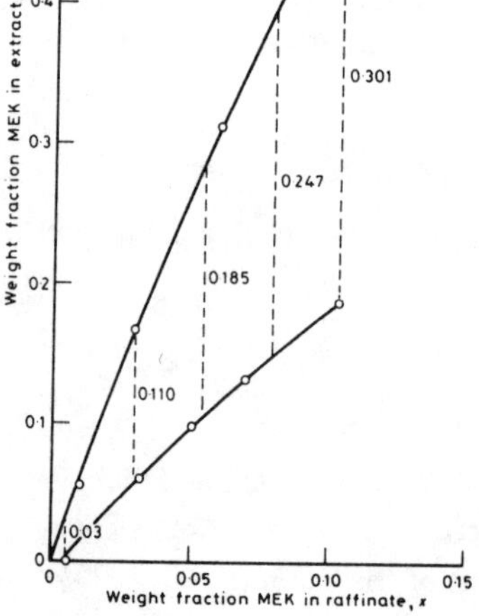

Figure 7.4 – *Driving force diagram for methyl ethyl ketone extraction.*

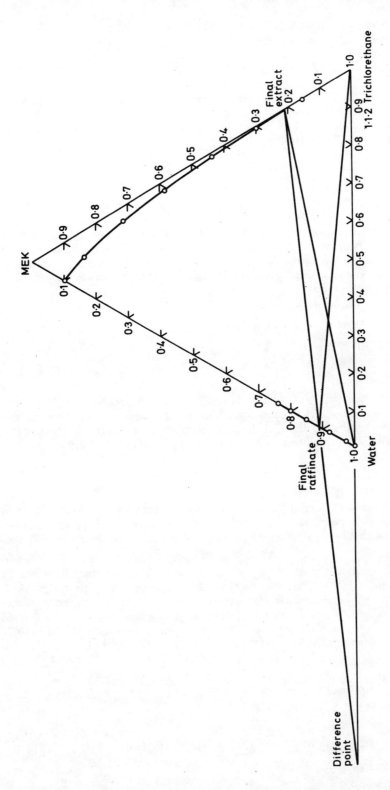

Figure 7.5 — *Graphical material balance on extraction.*

and the mean concentration gradient is

$$0.1774\rho_E = (C^* - C)_m$$

where $\rho_E = 1221.4$ kg/m³ is the mean density of the extract phase corresponding to a driving force of 0.1774

i.e. $(C^* - C)_m = 216.7$ kg/m³

7.4 Interfacial area

The interfacial area per unit height of column depends on the drop size in the dispersed phase and the fractional hold-up of the dispersed phase in the column. In order to evaluate these terms in an RDC it is necessary to specify the column diameter and the dimensions of the column internals. In Section 7.2 the column diameter was estimated to be 0.20 metres. The dimensions of the internals can then be estimated from the Chart prepared by Misek [35]. Thus for the column proposed.

Column Diameter, $D_C = 0.20$ metres
Rotor Diameter, $D_R = 0.10$ metres
Stator Diameter, $D_S = 0.15$ metres
Stator Spacing, $h = 0.05$ metres

These dimensions will be used to estimate drop-size and hold-up. The drop size depends on the method of drop formation and its subsequent treatment in the column by the rotating disc. This involves designing a distributor and assessing the rotor speed that will maintain the size of the drops formed at the distributor.

7.4.1 *Design of distributors*

The interfacial area depends on the drop size and residence time in the column. In this design it is proposed to disperse the solvent phase because it has the lower metric throughput and a dispersion with a narrow size distribution is more easily formed.

The interfacial tension of the system MEK/water/1.1.2 trichlorethane is moderately high (39.48×10^{-3} N/m) and therefore drops formed at a nozzle will be much larger than the nozzle diameter. Hence nozzles 0.2 cm will be recommended for the distributor. Then the area per nozzle = 0.0314 cm². If discrete drops are to be formed and jetting is to be avoided the maximum velocity through the nozzle will be 30 cm/s and the volumetric throughput per nozzle = 0.9425×10^{-6} m³/s.

$$\text{Number of nozzles required} = \frac{0.664}{0.9425 \times 10^{-6} \times 3600} = 195$$

∴ Propose 200 nozzles on a triangular pitch 0.4 cm. Area per nozzle = $2 \times 0.4 \times 0.2 = 0.16$ cm² and area per distributor = $(200 \times 0.16 \times 10^{-4})$ m²

Allowing a rim of 1.0 cm the diameter of the distributor will be 0.075 m. This will be satisfactory in the top of the column.

The velocity through each nozzle will be Q where

$$Q = \frac{0.664}{200 \times 3600} = 0.922 \times 10^{-6} \text{ m}^3/\text{s}$$

and

$$\text{Velocity through each nozzle, } U = \frac{0.922 \times 10^{-6}}{0.0314 \times 10^{-4}} = 0.294 \text{ m/s}$$

Then by Meister and Scheele[36]:—

$$V = F\left[\frac{\pi \theta D_N}{\Delta \rho g} + \frac{20.0 \mu_c Q D_N}{D_p^2 \Delta \rho g} - \frac{4.0 \rho_d Q U}{3.0 \Delta \rho g} + 4.5 \left(\frac{Q^2 D_N \rho_d \theta}{(\Delta \rho g)^2}\right)^{1/3}\right] \quad (7.10)$$

where

V is the drop volume

D_N is the nozzle diameter = 0.2×10^{-2} m

θ is the interfacial tension = 39.48×10^{-3} N/m

D_p is the drop diameter

The drag term involving the unknown D_p is negligible because the viscosity $\mu_c < 0.01$ kg/m s.

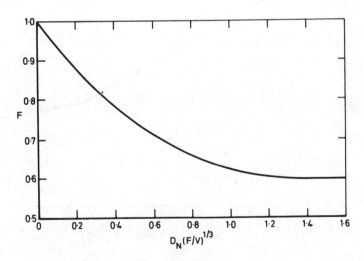

Figure 7.6 — *Harkin-Brown correction factor.*

$$V = F\left[\frac{\pi \times 39.48 \times 10^{-3} \times 0.2 \times 10^{-2}}{330 \times 9.81} - \frac{4.0 \times 1325 \times 0.922 \times 10^{-6} \times 0.294}{3.0 \times 330 \times 9.81}\right.$$
$$\left. + 4.5\left(\frac{(0.922 \times 10^{-6})^2 \times 0.2 \times 10^{-2} \times 1325 \times 39.48 \times 10^{-3}}{(330 \times 9.81)^2}\right)^{1/3}\right]$$
$$= F(0.0766 - 0.1480 + 0.1978)10^{-6} \text{ m}^3$$

or $(V/F) = 0.1264 \times 10^{-6}$ m³ and $(F/V)^{1/3} D_N = 0.3984$.

The Harkin-Brown Factor is from Figure 7.6 = 0.78 and the size of the drops formed at the distributor is: $D_P = 0.67$ cm

7.4.2 Design speed of rotating disc

The speed of the rotating disc that will maintain a drop size of 0.67 cm may be obtained from Kolmogroff's Law [3,7] which may be stated:

$$d_{max} = 0.72(\theta g/\rho_c)^{3/5} E^{-2/5} \qquad (7.11)$$

where E is power per unit mass.

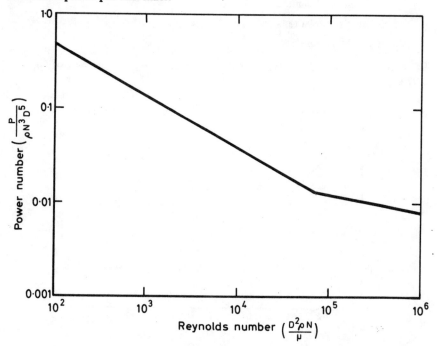

Figure 7.7 — *Relation between power number and Reynolds number for rotating disc contactor.*

Kung and Beckmann found that the mean drop size is 0.7 d_{max} so that [38]

$$d_m = 0.5(39.48g/1000g_c)^{0.6} E^{-0.4} = 0.67 \text{ cm}$$

or

$$E = 11.925 \text{ W/m}^3$$

Now

$$E = \frac{4.0 \times 10^3 C N^3 D_R^5}{\pi h D_c^2} \qquad (7.12)$$

where:—

C is a factor related to the disc Reynolds number as shown in Figure 7.7.

N is the angular velocity of the disc in revolutions per second.

That is

$$11.925 = \frac{4.0 \times 10^3 \times 0.10^5 \times N^3 \times C}{\pi \times 0.05 \times 0.2^2}$$

or

$$CN^3 = 1.8732 \qquad (7.13)$$

The Disc Reynolds Number is

$$Re = 0.1^2 \times 1000.0 \times N/0.00095 = 10526.32N \qquad (7.14)$$

From Figure 7.7, for $N = 5$ rps, $Re = 52631$ and $C = 0.012$. Then from equation (7.13)

$$N = \sqrt[3]{(1.8732/0.012)} = 5.38 \text{ rps}$$

The agreement is satisfactory and it will be recommended that the rotor rotates at 5 rps. This is equivalent to a rotor trip speed of

$$\pi \times 10.0 \times 300 = 94.25 \text{ metres/min.}$$

Kung and Beckmann[38] stated that the rotor tip speed should exceed 91.5 metres per minute to prevent the accumulation of drops beneath the disc. Again the proposed rotor speed is satisfactory. Finally Taylor vortices will be formed if the Taylor number exceeds 42.0. That is if:

$$T = \left(\frac{U D_c \rho_c}{\mu_c}\right) \left(\frac{D_c - D_R}{0.5 D_R}\right)^{0.5} \qquad (7.15)$$

where U is the linear velocity of the continuous phase. That is

$$U = \frac{2.2166}{0.7854 \times 0.2^2 \times 3600} = 0.0196 \text{ m/s}$$

and $$T = \left(\frac{0.0196 \times 1000.0 \times 0.20}{0.0008}\right)\left(\frac{0.2-0.1}{0.5 \times 0.1}\right)^{0.5} = 6929.$$

That is the rotor speed is sufficient to create Taylor vortices so that the RDC will function correctly.

7.4.3 Column hold-up
The fractional hold-up of the dispersed phase can be estimated from the relation

$$\frac{V_d}{X} + \frac{V_c}{1-X} = V_s(1-X) \tag{7.16}$$

where

V_d the superficial velocity of the dispersed phase
$= 0.664/(0.7854 \times 0.2^2) = 21.14$ m/h

V_c the superficial velocity of the continuous phase
$= 2.2166/(0.7854 \times 0.2^2) = 70.56$ m/h

X is the fraction hold-up of the dispersed phase.

V_s is the slip velocity, m/h

The slip velocity may be obtained from the correlation proposed by Kung and Beckmann; thus

$$\frac{V_s \mu_c}{\theta} = K\left(\frac{\Delta\rho}{\rho_c}\right)^{0.9}\left(\frac{g}{D_R N^2}\right)^{1.0}\left(\frac{D_S}{D_R}\right)^{2.3}\left(\frac{h}{D_R}\right)^{0.9}\left(\frac{D_R}{D_C}\right)^{2.6} \tag{7.17}$$

Since $(D_S - D_R)/D_C = (0.15 - 0.1)/0.2 = 0.25 > 0.04$; $K = 0.012$ and

$$\frac{V_s \times 0.00095}{39.48 \times 10^{-3}}$$

$$= 0.012\left(\frac{330}{1000.0}\right)^{0.9}\left(\frac{9.807}{0.1 \times 5.0^2}\right)^{1.0}\left(\frac{0.15}{0.10}\right)^{2.3}\left(\frac{0.05}{0.10}\right)^{0.9}\left(\frac{0.10}{0.20}\right)^{2.6}$$

or

$V_s = 0.4987 \times 0.3687 \times 3.9228 \times 2.541 \times 0.5359 \times 0.1649$

$= 0.1620$ m/s

$= 583.2$ m/h

Substituting into equation (7.16) gives

$$\frac{21.14}{X} + \frac{70.56}{1-X} = 583.2(1-X) \tag{7.18}$$

equation (7.18) can be rearranged to: $X^3 - 2X^2 - 0.9153X = 0.0362$ which can be solved by the method suggested by Jenson and Jeffreys[8]. Unfortunately there are two real roots less than 1.0. That is: $X = 0.622$ and $X = 0.045$.

However, the slip velocity has been found to be 0.162 m/s, suggesting that the residence time of the dispersed phase drop in a compartment 0.05 m high will be 0.3 s. This is a very approximate estimate but it does indicate that $X = 0.045$ is the more realistic value of the hold-up in the RDC proposed here. Furthermore, checking the hold-up using the method of Misek[35] confirms that the hold-up is $X = 0.045$. On this basis the interfacial area per compartment of the proposed RDC is:

Volume of dispersed phase per compartment
$= 0.045 \times \frac{1}{4} \pi \times 0.20^2 \times 0.05 = 70.69 \times 10^6$ m^3

Drop volume $= \frac{1}{6} \pi \times (0.67 \times 10^{-2})^3 = 0.157 \times 10^{-5}$ m^3

Number of drops per Compartment
$= (70.69 \times 10^6)/(0.1575 \times 10^{-6}) = 449$ drops.

and the interfacial area is

$$\pi \times (0.67 \times 10^{-2})^2 \times 449 = 0.0633 \text{ m}^2/\text{compartment}$$

7.5 Mass transfer coefficient

The overall mass transfer coefficient in an extraction process is related to the continuous phase mass transfer coefficient and the dispersed phase mass transfer coefficient by the relation:

$$\frac{1}{K_G} = \frac{1}{k_d} + \frac{1}{mk_c} \qquad (7.19)$$

where:-

k_d is the dispersed phase mass transfer coefficient, m/s

k_c is the continuous phase mass transfer coefficient, m/s

m is the distribution ratio = 5.7

Each of the coefficients will be evaluated.

7.5.1 *The dispersed phase mass transfer coefficient*

The dispersed phase mass transfer coefficient depends on the hydrodynamic behaviour of the drops. That is, whether the liquid inside the drop is stagnant, circulating or whether the drops are oscillating. The drops are 0.67 cm in diameter and the drop Reynolds number is

$$= \frac{0.0067 \times 0.162 \times 1325.0}{0.00095} = 1513$$

and therefore the drops are oscillating. That is the drop Reynolds number exceeds 200. Then by Rose and Kintner[39] the dispersed phase coefficient is

$$k_d = 0.45(D_d \omega)^{0.5} \qquad (7.20)$$

where

D_d is the diffusivity of MEK in 1.1.2 trichlorethane
ω is the frequency of drop oscillation.

The diffusivity can be estimated by the Wilke-Chang correlation

$$D_d = \frac{7.48 \times 10^{-12} T}{\mu} \frac{(\psi M_B)^{0.5}}{V^{0.6}} \text{ in m}^2/\text{s} \qquad (7.21)$$

where

V is molar volume (From Perry[19]).
For MEK = $(4 \times 14.8) + (8 \times 3.7) + 7.4 = 96.2 \text{ cm}^3/\text{gm mole}$

T is temperature = 300 K

M_B is molecular weight, 18 for water 133.4 for trichlorethane.

ψ is association parameter, 2.6 for water 1.0 for trichlorethane.

and

$$D_{MEK} = \frac{7.48 \times 10^{-12} \times 300 (2.6 \times 18.0)^{0.5}}{0.95 \times 92.6^{0.6}}$$

$$= 1.03 \times 10^{-9} \text{ m}^2/\text{s in water}$$
$$= 2.06 \times 10^{-9} \text{ m}^2/\text{s in trichlorethane}$$

Rose and Kintner expressed ω in terms of the drop size and the physical properties of the system by

$$\omega^2 = \frac{\sigma b}{r^3}\left[\frac{n(n-1)(n+1)(n+2)}{(n-1)\rho_d + n_c}\right] \qquad (7.22)$$

where

r is drop radius in cm = 0.335 cm

n is an integer = 2.0

b is a constant depending on drop size = $d^{0.225}/1.242$ where d is the drop size in centimetres.

Then
$$\omega^2 = \frac{39.48 \times 0.67^{0.225}}{1.242 \times 0.335^3} \left(\frac{1 \times 2 \times 3 \times 4}{3 \times 1.325 + 2 \times 1.0} \right) = 3107$$

or $\omega = 55.7/\text{s}$ and $k_d = 0.45 \, [2.06 \times 10^{-9} \times 55.7]^{0.5} = 0.000152$ m/s.

7.5.2 *The continuous phase mass transfer coefficient*

It is generally agreed that the continuous phase mass transfer coefficient is best estimated from the correlation proposed by Garner, Foord and Tayeban [40]. This is expressed

$$k_c d/D = -126. + 1.8 Re^{0.5} Sc^{0.42} \qquad (7.23)$$

where

Re is the drop Reynolds number = 1513
Sc is the Schmidt number

$$= \left(\frac{0.00095}{1000 \times 1.03 \times 10^{-9}} \right)^{0.42} = 17.59$$

Then
$$k_c(0.67 \times 10^{-2})/(1.03 \times 10^{-9}) = -126 + (1.8 \times 38.89 \times 17.59)$$

or
$$k_c = 0.000169 \text{ m/s}$$

and the overall mass transfer coefficient is, from equation (7.19).

$$\frac{1}{K_G} = \frac{1}{0.000152} + \frac{1}{5.7 \times 0.000169}$$
$$= 6579 + 1038 = 7617 \text{ s/m}$$
$$K_G = 0.000131 \text{ m/s}$$

7.6 Height of extraction column

The rate of mass transfer necessary to extract 227.2 kg/h of MEK and 2-butanol is $227.2 = 0.000131 \times 0.0633 N \times 1221.4 \times 0.1774 \times 3600$ where N is the number of compartments required, = 35.13, say 36. Then the height required is 1.80 m, say 2.00 m.

The interfacial tension of the system is 39.48 N/m which is relatively high and therefore the rate of coalescence of the dispersed phase emerging from the bottom compartment will be high. However a space of height 0.30 m will be allowed.

Finally 0.20 m will be allowed to accommodate the distributor so that the total height of the extraction column will be 2.50 m.

The final specification of the extraction column is

Column height = 2.50 m

Column diameter = 0.2 m

Rotor diameter = 0.1 m

Stator diameter = 0.15 m

Stator spacing = 0.05 m

Distributor diameter = 0.075 m

Diameter of holes in distributor = 0.002 m

Number of holes in distributor = 200

7.7 Recovery of MEK 2-butanol from extract

The extract phase with the composition stated in Table 7.1 will be discharged to a solvent recovery still at the rate of 1126.91 kg/h. Examination of the extract column of Table 7.1 shows that the extract contains 0.14% water and 1.24% 2-butanol so that this system contains four components. However the amount of water and 2-butanol present is so small that these will not be able to exert any affect on the transport processes occurring in the solvent recovery still. Therefore the design of this unit will be based on the binary mixture MEK/trichlorethane and the equilibrium diagram for this binary was evaluated from Raoult's law and is presented in Table 7.2. Examination of this table shows that MEK is considerably more volatile than the trichlorethane so that it should be possible to recover nearly all the solvent from the extract in a relatively small distillation unit.

7.7.1 *Material balance on solvent recovery still*

On the basis of the relative volatilities of MEK and the solvent, and the statement in the above section, the material balance on the solvent recovery unit is presented in Table 7.3. In this table the mass and kg mole flowrates are presented and it is obvious that a column of small diameter, of the order of 0.2 m will suffice. Hence a packed column will be recommended, but since the detailed designed of a packed column has already been described in Chapter 6 only the results of the calculations will be reported in this section. The procedures followed will be the same as those given in Chapter 6

Table 7.2 — *Vapour/liquid equilibrium data for the system methyl ethyl ketone/ 1.1.2 trichlorethane.*

Temperature	Mole fraction MEK in liquid	Mole fraction MEK in vapour
112.0	0.04	0.217
107.1	0.085	0.380
100.2	0.161	0.560
95.6	0.234	0.660
92.3	0.308	0.724
89.7	0.369	0.779
86.4	0.475	0.819
84.5	0.560	0.847
83.2	0.631	0.867
82.0	0.714	0.885
80.9	0.848	0.920
80.4	0.928	0.952

7.7.2 Analysis of feed conditions

The extract leaves the bottom of the extraction column at 27°C at the rate of 1126.11 kg/h with the molar composition stated in Table 7.3. Interpolation of this composition into Table 7.2 shows that the bubble point of the feed is 93°C and the heat required to raise the temperature of the feed to its bubble point is:-

1. MEK: $0.2968 \times 165.726 \times (93-27) = 3246.38$ (kJ)

2. 2-butanol: $0.019 \times 170.058 \times (93-27) = 213.25$ (kJ)

3. Water: $0.0087 \times 75.36 \times (93-27) = 43.27$ (kJ)

4. Trichlorethane: $0.6755 \times 148.56 \times (93-27) = 6623.25$ (kJ)

Heat required per kg mole of Feed: 10126.15 kJ/kg mole and Heat required to vaporise feed at its bubble point

1. MEK: $0.2968 \times 31696.17 = 9407.40$ kJ

2. 2-butanol: $0.019 \times 42825.10 = 813.58$ kJ

3. Water: $0.0087 \times 40695.70 = 354.04$ kJ

4. Trichlorethane: $0.6755 \times 31835.59 = 21504.95$ kJ

Latent heat of vaporisation per kg mole = 32079.97 kJ

Table 7.3 – *Material balance on solvent recovery still.*

Component	Feed				Distillate				Bottom product			
	Weight rate (kg/h)	(%)	kg Mole rate (kg/h)	(%)	Weight rate (kg/h)	(%)	kg mole rate (kg/h)	(%)	Weight rate (kg/h)	(%)	kg mole rate (kg/h)	(%)
MEK	214.40	19.00	2.957	29.68	212.00	93.75	2.940	93.93	1.20	0.14	0.017	0.25
2-butanol	14.00	1.24	0.189	1.90	14.00	6.19	0.189	6.04	–	–	–	–
Water	1.57	0.14	0.087	0.87	–	–	–	–	1.57	0.17	0.087	1.27
Trichlorethane	896.14	79.62	6.731	67.55	0.14	0.06	0.001	0.03	898.00	99.69	6.730	98.48
Total	1126.11	100.00	9.964	100.0	226.14	100.0	3.13	100.00	900.77	100.00	6.834	100.00

Then
$$q = (32079.97 + 10126.15)/32079.97 = 1.32$$

In the above table on the evaluation of the latent heat of the feed it should be noticed that the molar latent heats of MEK and the trichlorethane are sufficiently close to justify the application of the McCabe-Thiele construction for the evaluation of the height of packing required. In addition the heat required to raise the temperature of the feed is small and does not justify the installation of a feed heater. That is cold feed will be pumped from the bottom of the extraction column into the solvent recovery still.

7.7.3 *Estimation of reflux ratio*

The equilibrium data presented in Table 7.2 is plotted in Figure 7.8 together with the distillate and bottom product composition. The slope of the feed condition line is
$$q/(q-1) = 1.32/(1.32-1.0) = 4.13$$

and is shown on Figure 7.8. In addition the operating line for the rectifying section at minimum reflux ratio is also shown. From the intercept of this line $0.9393 = 0.68 (R_m + 1)$ or the minimum reflux ratio is 0.38. The optimum reflux ratio is

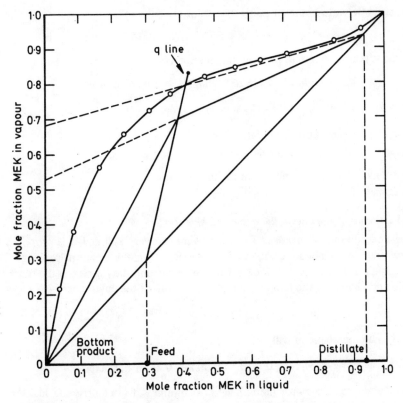

Figure 7.8 – *McCabe-Thiele diagram of solvent recovery unit.*

approximately $2.0\ R_m$ and this relation will be accepted in this design report. That is $R = 0.76:1.0$ will be recommended. Then the equation of the rectifying section operating line will be

$$y = 0.43x + 0.53 \qquad (7.24)$$

Equation (7.24) is drawn on Figure 7.8 and the McCabe-Thiele construction is completed as shown in that figure.

7.7.4 Vapour and liquid flow rates through column

The feed, distillate and bottom product flowrates are presented in Table 7.3. From these rates the vapour flowrate in the rectifying section of the column is

$$V_n = D(R+1) = 3.13(0.76+1) = 5.509 \text{ kg moles/h}$$
$$\equiv 0.0483 \text{ m}^3/\text{s}$$

The liquid flowrate in the rectifying section is

$$RD = 0.76 \times 3.13 = 2.38 \text{ kg moles/h}$$

The liquid flowrate in the stripping section of the column is

$$O_m = O_n + qF = 2.38 + 1.32 \times 9.964 = 15.531 \text{ kg moles/h}$$

and the vapour flowrate in the stripping section is

$$V_m = O_m - W = (15.531 - 6.834) = 8.697 \text{ kg moles/h}$$

At the feed plate (0.32×9.964) kg moles/h condense.

Then vapour rising above feed plate

$$V_n = (8.697 - 3.188) = 5.509 \text{ kg moles/h}$$

confirming the value obtained from the reflux ratio.

7.7.5 Estimation of column diameter and height

A packed column is envisaged for this solvent recovery still and it is proposed to utilise 2.5 cm diameter stoneware Raschig rings for the column packing. The economic gas rate for this packing is $1825 \text{ m}^3/\text{h m}^2$ and the gas rate in the stripping section and based on the temperature at the bottom of the column is

$$8.697 \times 22.4 \times 385/273 = 274.74 \text{ m}^3/\text{h}$$

and the column diameter will be

$$274.74/1825.0 = 0.1505 \text{ m}^2 \equiv 0.44 \text{ m diameter}$$

then, the interfacial area per metre height of column is, from Morris and Jackson [31].

$$(184.0 \times 0.1505) = 27.7 \text{ m}^2/\text{m height of column}$$

The vapour velocity based on the stripping section is

$$U_{av} = 0.0763/(0.1505 \times 0.8) = 0.63 \text{ m/s}$$

where   The packing porosity is 0.8
Viscosity of vapour = 8.34×10^{-6} kg/m s
Vapour density = 2.48 kg/m³

and   Reynolds Number $= \dfrac{0.025 \times 2.48 \times 0.63}{8.34 \times 10^{-6}} = 4762$

Diffusivity of MEK vapour in 2-butanol $= 6.006 \times 10^{-6}$ m²/s

Schmidt Number $= \dfrac{8.34 \times 10^{-6}}{2.48 \times 6.006 \times 10^{-6}} = 0.56$

In the solvent recovery still the mass transfer rate will be gas phase controlled and the mass transfer coefficient must be based on equi-molar counter-current diffusion processes. For this type of process and for a Schmidt number of 0.6 the mass transfer coefficient can be estimated from the correlation

$$j_D(10^{-1}) = (K_y/G_m)Sc^{0.67}(10^{-1}) \tag{7.25}$$

For a Reynolds number of 4762, $j_D (10^{-1}) = 0.0033$ (Treybal[41]) and

$$G_m = 8.697/(0.1505 \times 3600) = 0.016 \text{ kg moles/m}^2 \text{ s}$$

and the mass transfer coefficient is

$$K_y = 0.0033 \times 10 \times 0.016/(0.56^{0.67}) = 7.8 \times 10^{-4} \text{ kg moles/m}^2 \text{ s}$$

The height of the solvent recovery column is

$$Z = \dfrac{G}{K_y a} \int_{y_2}^{y_1} \dfrac{dy}{y^* - y} \tag{7.26}$$

The integral in equation (7.26) can be evaluated from the McCabe-Thiele diagram, Figure 7.8 by abstracting the appropriate values of y and y^* as shown in Table 7.4.

From Table 7.4 the integral in equation (7.26) is

$$\int \dfrac{dy}{y^* - y} = \left(\dfrac{0.9393 - 0.0025}{36}\right) [(48.309 + 21.0526)$$

$$+ 4(29.5858 + 11.1235 + 5.8480 + 4.0290 + 3.7273 + 6.2735)$$

$$+ 2(17.6058 + 7.8125 + 4.7148 + 3.6873 + 4.0917)]$$

$$= 9.839$$

and

$$Z = (0.016 \times 9.839)/(7.8 \times 10^{-4} \times 27.7) = 7.3 \text{ m}$$

Table 7.4 – *Data for evaluation of integral in equation (7.26).*

y	y^*	y^*-y	$1/(y^*-y)$
0.9393	0.960	0.0207	48.309
0.8612	0.895	0.0338	29.5858
0.7832	0.840	0.0568	17.6058
0.7051	0.795	0.0899	11.1235
0.6270	0.755	0.1280	7.8125
0.5490	0.720	0.1710	5.8480
0.4709	0.683	0.2120	4.7148
0.3928	0.641	0.2482	4.0290
0.3148	0.586	0.2712	3.6873
0.2367	0.505	0.2683	3.7272
0.1586	0.403	0.2444	4.0917
0.0806	0.240	0.1594	6.2735
0.0025	0.050	0.0475	21.0526

7.7.6 Solvent recovery still reboiler

The detailed design of a thermo-syphon reboiler was presented in section 4.10 and will not be repeated here. A summary of the design calculations is as follows. Heat Duty: 278843.2 kJ/h. The reboiler is heated with steam at a pressure of 2.0 bar. Steam consumption 130 kg/h. Heat transfer area required = 2.05 m^2

7.7.7 Solvent recovery still condenser

The detailed design of a condenser is presented in section 8.4 and the calculations will not be repeated here. A summary of the calculations is:-

Heat duty = 179508.2 kJ/h

Condensing temperature = 80°C

Log mean temperature difference = 41.7°C

Heat transfer coefficient = 850 W/m^2 K

Heat transfer area = 1.45 metres^2

Water requirement = 16.50 kg/h

7.7.8 Bottom product cooler

The trichlorethane discharged from the bottom of the solvent recovery still at the rate of 900.77 kg/h and at a temperature of 112°C is to be cooled to 27°C before recycling to the solvent extraction column. A summary of the design of this cooler is presented.

$\quad$ Heat duty = 85801.6 kJ/h

$\quad$ Log mean temperature difference = 19.5°C

$\quad$ Heat transfer coefficient = 570 W/m² K

$\quad$ Heat transfer area = 2.25 metres²

$\quad$ Water requirement = 790 kg/h

7.7.9 Product still intermediate storage

The distillate from the solvent recovery still leaves the condenser at 80°C and will be made to pass into a storage tank receiving the condensate from the cooler-condenser at 27°C. The temperature of the resulting mixture is estimated to be

1. Heat Content of Solvent Recovery Still Distillate

$\quad$ (i) MEK: 212.0 × 2.2986 $\quad$ = 487.29 kJ/h°C
$\quad$ (ii) Alcohol: 14.0 × 2.4367 $\quad$ = 34.11
$\quad$ (iii) Water: 0.14 × 1.11 $\quad\quad\quad$ = 0.16
$\quad\quad\quad\quad\quad\quad\quad\quad$ Total $\quad$ = 521.56 kJ/h°C

2. Heat Content of Cooler-Condenser Condensate

$\quad$ (i) MEK: 1133.6 × 2.2986 $\quad$ = 2605.69 kJ/h°C
$\quad$ (ii) Alcohol: 155.24 × 2.4367 = 378.27
$\quad\quad\quad\quad\quad\quad\quad\quad$ Total $\quad$ = 2983.96 kJ/h°C

Then 521.56 (80−t) = 2983.96 (t−27) or temperature of feed to product still = 35°C. This concludes the design of the Extraction Column.

Chapter 8

THE PRODUCT DISTILLATION UNIT

8.1 General discussion

The product distillation unit will be fed with a MEK/2-butanol mixture obtained from the condensate from the cooler-condenser and from the distillate of the solvent recovery still of the extraction plant. These two streams are to be mixed in an intermediate storage holding tank and then fed to the product distillation unit. The temperature of this feed was shown to be 35°C in Section 7.7.9 and it will consist of 1345.6 kg/h of MEK and 169.24 kg/h of 2-butanol, with a trace of trichlorethane. This mixture is to be distilled to produce a top product containing 99.0% weight MEK and a bottom product containing 99% weight 2-butanol. The 2-butanol will be recycled and mixed with the feed to the reactor. For these conditions the material balance over the product distillation unit is presented in Table 8.1, expressed in flowrates of kg/h and kg mole/h.

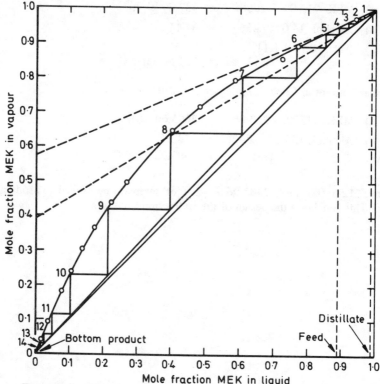

Figure 8.1 – *McCabe-Thiele diagram for product distillation unit.*

Table 8.1 – *Material balance on product distillation unit.*

Component	Feed				Distillate				Bottom product			
	Weight rate		kg mole rate		Weight rate		kg mole rate		Weight rate		kg mole rate	
	(kg/h)	(%)	(kg mole/h)	(%)	(kg/h)	(%)	(kg mole/h)	(%)	(kg/h)	(%)	(kg mole/h)	(%)
MEK	1345.60	88.82	18.663	89.10	1344.03	99.00	18.641	99.03	1.57	0.92	0.022	1.04
2-Butanol	169.24	11.17	2.283	10.89	13.58	1.00	0.183	0.97	155.66	99.00	2.100	98.91
Trichlorethane	0.14	0.01	0.001	0.01					0.14	0.08	0.001	0.05
Total	1514.98	100.00	20.947	100.00	1357.61	100.00	18.824	100.00	157.37	100.00	2.123	100.00

8.2 Selection of column

The equilibrium data for the system 2-butanol/MEK has been published by Amick, Weiss, Kirshenbaum[42] and their data are presented in Figure 8.1 together with the compositions of the feed, distillate and bottom product. This diagram emphasises that the product distillation column is essentially a stripping column with very few plates, or packing, above the feed point. Furthermore, the feed flowrate is 1514.98 kg/h, which is sufficiently large to suggest that the feed should be preheated prior to entering the column. This will be proposed and a feed heater capable of heating the feed to its boiling point will be installed in this unit of the plant. On this basis the "q line" will be vertical and the rectifying operating line corresponding to the minimum reflux ratio is shown dotted in Figure 8.1. From this line

$$x_D/(R_m+1) = 0.99/(R_m+1) = 0.572$$

That is the minimum reflux ratio = 0.73. Following Coulson and Richardson[10] the optimum reflux ratio = $2R_m$ = 1.46, say 1.5. An operating reflux ratio of 1.5 will be recommended.

The vapour rate will then be

$$V = D(R+1) = 2.5 \times 18.824 = 47.06 \text{ kg moles/h}$$

Since the feed enters at its boiling point this vapour rate will be constant throughout the column. However, the temperature at the bottom of the column will be 99°C and the volumetric flowrate will be

$$(47.06 \times 22.4 \times 372.0)/273.0 = 1436.42 \text{ m}^3/\text{h} = 0.40 \text{ m}^3/\text{s}$$

8.2.1 *Minimum column diameter*

If the column is to be a plate column the minimum column diameter will be found as follows:

Treybal states that the superficial gas velocity in a plate column can be estimated from the relation.

$$V_F = C_F \left(\frac{\rho_L - \rho_g}{\rho_g}\right)^{0.5} \quad (8.1)$$

where V_F is the flooding velocity based on the bubbling area

C_F is a coefficient obtainable from

$$C_F = \left[a \log\left(\frac{1}{(L/G)(\rho_g/\rho_L)^{0.5}}\right) + b\right] \left(\frac{\gamma}{20}\right)^{0.2} \quad (8.2)$$

where a and b are coefficients dependent on the parameter $(L/G)(\rho_g/\rho_L)^{0.5}$ and the plate spacing.

For the product distillation column $G = 3488.1$ kg/h, $L = 3645.4$ kg/h, then

$$\frac{L}{G}\left(\frac{\rho_g}{\rho_L}\right)^{0.5} = \frac{3488.1}{3645.4}\left(\frac{2.4283}{804.0}\right)^{0.5} = 0.0526$$

and from Table 6.1 in Treybal[41] $a = 0.0873$ and $b = 0.1535$ for a plate spacing of 0.46 m. The minimum recommended for columns of the order of 1.0 m diameter. Then

$$C_F = \left[0.0873 \log\left(\frac{1.0}{0.0526}\right) + 0.1535\right]\left(\frac{31.2}{20.0}\right)^{0.2}$$
$$= 0.2898 \text{ ft/s} \equiv 0.0883 \text{ m/s}$$

and

$$V_F = 0.0883\left(\frac{804 - 2.4283}{2.4283}\right)^{0.5} = 1.60 \text{ m/s}$$

That is the maximum vapour velocity would be 1.60 m/s which passes through the bubbling area of the plate.

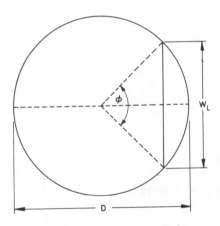

Figure 8.2 — *Calculation of weir length.*

The bubbling area may be estimated as follows. From Figure 8.2 the downcomer area is

$$\frac{\pi}{4}\left(\frac{\phi}{360}\right)D^2 - \left(\frac{W_l}{2} \frac{D}{2} \cos\frac{\phi}{2}\right) \quad (8.3)$$

and for a column that is to operate at atmospheric pressure Treybal states [41] that $W_1 = 0.7\,D$ Therefore $\sin(\phi/2) = 0.35D/0.5D = 0.7$ or $\phi = 89.0°$ Then the downcomer area

$$= \left[\left(\frac{0.7854 \times 89}{360}\right) - \left(\frac{0.7 \times 0.7133}{4}\right)\right] D^2$$

$$= 0.07 D^2$$

so that the bubbling area $= 0.7854 D^2 - 0.14 D^2 = 0.6454 D^2$.

Now the net plate area will be $0.40/1.60 = 0.25$ m^2

and the minimum column diameter is

$$D = (0.250/0.6454)^{0.5} = 0.622 \text{ m}$$

However to ensure that the column would not flood, a column 1.0 m in diameter will be recommended.

8.2.2 Confirmation of recommended column diameter

A plate column of diameter 1.0 m has been proposed and in this section its stability during operation will be checked. Thus Treybal[41] states that the minimum vapour velocity through the holes in a sieve plate necessary to prevent weeping can be estimated by the empirical correlation

$$\frac{V_w \mu_g}{\theta} = 2.92 \times 10^{-4} \left(\frac{\mu_g^2 \rho_L \times 10^5}{\theta \rho_G^2 d_h}\right)^{0.379} \times \left(\frac{l}{d_h}\right)^{0.293} \left(\frac{2 A d_h}{\sqrt{3 p^3}}\right) \quad (8.4)$$

where

$$\zeta = \frac{2.8}{(Z/d_h)^{0.724}}$$

V_w is weeping velocity, m/s
μ_g is 0.48×10^{-5} kg/m s
θ is 31.2×10^{-3} kg/s^2
d_h is diameter of hole in sieve plate $= 0.0047$ m
p is pitch between holes (triangular) $= 0.0127$ m
Z is length of bubbling area $= 0.7133$ m
A is area of bubbling section $= 0.645$ m^2
l is thickness of metal in plate $= 0.002$ m

Then $(Z/d_h)^{0.724} = (0.7133/0.0047)^{0.724} = 37.95$ and $\zeta = 2.8/37.95 = 0.0738$.

The minimum velocity through the holes is

$$V_w = \left(\frac{2.92 \times 10^{-4} \times 31.2 \times 10^{-3}}{0.48 \times 10^{-5}}\right) \left(\frac{(0.48 \times 10^{-5})^2 \times 331 \times 10^5}{31.2 \times 10^{-3} \times 2.4283 \times 0.0047}\right)^{0.379}$$

$$\times \left(\frac{0.002}{0.0047}\right)^{0.293} \left(\frac{2.0 \times 0.645 \times 0.0047}{\sqrt{3} \times 0.0127^3}\right)^{0.0738}$$

$$= 1.898 \times 1.335 \times 0.778 \times 1.732 = 3.41 \text{ m/s}$$

The design velocity is found as follows.

Sieve plate holes are to be arranged on a triangular pitch of 0.0127 m,
Then the area per hole = 0.00014 m²

Number of holes = (0.645 − 0.028 clearance)/0.00014 = 4400 holes.

Total area of holes = $\frac{1}{4} \pi \times 0.0047^2 \times 4400 = 0.0763$ m²
and the velocity through the holes = 0.40/0.0763 = 5.24 m/s.

In addition the holes occupy 12.0% of the bubbling area of the plate. This is well within the recommended limits and will not necessitate adjusting C_F. The actual superficial velocity will be = 0.40/0.645 = 0.62 m/s, and $V/V_F = 0.62/1.60 = 0.39$. Then from Treybal's Figure 6.11 the fractional entrainment will be = 0.006. i.e. 0.6% which is negligible.

Let the head over the crest of the weir be 1.5 cm = h_1 and then $h_1/D = 0.6/3.28 = 0.18$, and from Treybal's Figure 6.10, $(W/W_{eff} = 1.04$ or $W_{eff} = 0.66$ m. Then

$$h_1 = 13.67 \left(\frac{W}{W_{eff}}\right)^{2/3} \left(\frac{q}{W}\right)^{2/3}$$

$$= 13.67 \times 1.04 \times \left(\frac{0.0445}{2.296}\right)^{2/3}$$

$$= 1.03 \text{ cm which is acceptable.}$$

The weir height h_w will be set at 5.0 cm. Then the orifice coefficient will be

$$C_0 = 1.09 \left(\frac{0.0047}{0.002}\right)^{0.25} = 1.35$$

The hole Reynold's number is

$$= \frac{0.0047 \times 2.4283 \times 5.24}{0.48 \times 10^{-5}} = 1.25 \times 10^4$$

Then $f = 0.0085$ (Perry Figure 5-26) and

$$h_D = \frac{C_0 V_h^2 \rho_G}{2g\rho_L}\left[0.4\left(1.25 - \frac{A_h}{A_n}\right) + \frac{4lf}{d_h} + \left(1.0 - \frac{A_h}{A_n}\right)^2\right] \tag{8.6}$$

$$= \frac{1.35 \times 5.24^2 \times 2.4283}{9.808 \times 804 \times 2}\left[0.4\left(1.25 - \frac{0.0763}{0.645}\right)\right.$$

$$\left. + \frac{4.0 \times 0.002 \times 0.0085}{0.0047} + \left(1.0 - \frac{0.0763}{0.645}\right)^2\right]$$

$$= 0.0014(0.353 + 0.0145 + 0.774) = 0.013 \text{ m}$$

$$= 1.3 \text{ cm liquid.}$$

Hydraulic head, h_L is estimated from the empirical correlation

$$h_L = 0.24 + 0.725h_w + 0.29h_w V\rho_G^{0.5} + 4.48q/Z \tag{8.7}$$

where the dimensions are in British Units, i.e.

$$= 0.24 + 0.725 \times 2.0 + 0.29 \times 2.0 \times 2.03 \times 0.389$$
$$+ (4.48 \times 0.0445)/2.789 \text{ inches}$$
$$= 3.3 \text{ cm}$$

The Residual Pressure Drop h_R may be estimated from the empirical correlation

$$h_R = 0.06\sigma/\rho_L d_R \qquad (8.8)$$

where σ is surface tension in dynes
  p is liquid density in British units
  d_R is bubble diameter in British units

$$h_R = \frac{0.06 \times 31.2 \times 2.54}{0.804 \times 62.5 \times 0.1875} = 0.5 \text{ cm}$$

Then the total gas pressure drop

$$h_G = 1.30 + 3.30 + 0.5 = 5.1 \text{ cm liquid}$$

Pressure Loss at Liquid Entrance h_2: The downspout apron will be set at $(h_w - 1.25)$ cm = 3.75 cm above the plate and the flow area beneath the apron will be 3.75 × 0.70/100 = 0.0263 cm² which is less than the downcomer area so that

$$h_2 = \frac{3}{2g}\left(\frac{q}{A da}\right)^2 = \frac{3}{2 \times 9.808}\left(\frac{4.534}{3600 \times 0.0263}\right)^2 = 0.035 \text{ cm}$$

and the "back-up" in the downcomer will be

$$h_3 = 5.10 + 0.04 = 5.14 \text{ cm}$$

Check on Flooding:

$$h_w + h_1 + h_3 = 5.0 + 1.03 + 5.14 = 11.2 \text{ cm}$$

This is considerably less than half the plate spacing or 23.0 cm. The plate will be stable in operation and the design is satsifactory.

Finally the downcomer residence time will be found:

Volume of downcomer = 0.007 (0.46 + 0.05) = 0.0357 m³

Liquid flowrate = 0.0013 m³/s

Residence time = 27 s.

This should be satisfactory.

This concludes the design of the sieve plate.

8.2.3 Estimation of number of plates required

The McCabe-Thiele diagram has been drawn in Figure 8.1 and this shows that 13 ideal plates will be required for the separation specified. The plate efficiency will be based on the bottom plate in the column.

8.2.4 Plate efficiency

The plate efficiency will be estimated by the *Bubble Tray Manual* published by the American Institute of Chemical Engineers. This prediction procedure is now well established and the background for the correlations and their limitations are discussed in the text of the manual and will not be considered here. Here it is proposed to estimate the tray efficiency by following the example appended to the manual without explanation. Furthermore, many of the correlations are expressed in a variety of dimensions, *e.g.* US gallons, and it has been decided to follow the same dimensions as those applied in the manual. Then following the format of the *Bubble Tray Manual* the estimation of the tray efficiency is performed as follows:

1. Slope of Equilibrium curve in the vicinity of the bottom tray = 2.5
2. Gas Rate = 0.62 m/s ≡ 2.03 ft/s = U_g
3. F-factor = $U_g p_g^{1/2}$ = 2.03 × $0.1516^{0.5}$ = 7.09. This value of the F-factor is only slightly below the lower limit recommended in the manual.
4. Liquid Rate = 3645.1 kg/h ≡ 19.94 US gal/min

$$\equiv \frac{19.94}{2.79} = 7.15 \text{US gal/min. width of tray}$$

5. $\lambda = mG/L$ where G = 47.06 kg mole/h, L = 49.18 kg mole/h and
   λ = 2.5 × 47.06/49.18 = 2.39.
6. $Z_f = 2.53F^2 + 1.89W - 1.6 = (2.53 \times 0.79^2) + (1.89 \times 2.0) - 1.6 = 3.76$
7. $Z_c = (1/\rho_L)(103. + 11.8W - 40.5F + 1.25L)$
   = $(1/50.25) [103. + (11.8 \times 2.0) - (40.5 \times 0.79) + (1.25 \times 7.15)]$
   ≡ 3.334 inches
8. $t_1 = 37.4 \, AZ_c/Q$
   where Q = 19.94 US gallons [see (4) above]
   A = 0.645 m² ≡ 6.95 ft²
   and t_1 = 37.4 × 6.95 × 3.334/19.94 = 43.4 s.
9. $N_1 = 103 \, D_1 (0.26F + 0.15) \, t_1$
   where
   $$D_l = \frac{2.87 \times 10^{-7} (XM)^{0.5} T}{V_{AB}^{0.6} \mu}$$
   and T = 373 K, μ = 0.58 cP, V_{AB} = 87.7
   so that
   $$D_l = \frac{28.7 \times 10^{-7} (1.0 \times 72.1)^{0.5} \times 373}{87.7^{0.6} \times 0.58}$$
   $$= 10.7 \times 10^{-5} \text{ ft}^2/\text{h}$$

and $\quad N_l = 103 \times 0.0103[(0.26 \times 0.79) + 0.15]43.4 = 16.4$

10. $N_g = (1.0/Sc^{0.5})\,[0.776+0.116W-0.29F+0.0217L]$
    $= (1.0/0.24^{0.5})[0.776+(0.116 \times 2.0)-(0.29 \times 0.79)+0.0217 \times 7.15]$
    $= 1.94$

11. $-\log(1-E_{oG}) = \dfrac{0.434 N_l N_g}{(N_l + \lambda N_g)}$

    $= \dfrac{0.434 \times 16.4 \times 1.94}{16.4+(2.39 \times 1.94)} = 0.656$

12. $E_{OG} = 0.74$

13. Percentage Resistance in Liquid $= \lambda N_g\, 100/(N_l + \lambda N_g)$

    $= \dfrac{2.39 \times 1.94 \times 100}{16.4+(2.39 \times 1.94)} = 6.09\%$

14. Eddy Diffusivity

    $D_E = [1.0+0.44(d-0.33)]^2 (0.0124+0.015W+0.0171U_g+0.0025L)^2$
    $= 1.0[0.0124+(0.015 \times 2.0)+(0.0171 \times 2.03)+(0.0025 \times 7.15)]^2$
    $= 0.009 \text{ ft}^2/\text{s}$

15. The Peclet number $Pe = Z_1^2/D_E t_1 = 2.34/(0.009 \times 43.4) = 6.02$.

16. $\lambda E_{og} = 2.39 \times 0.74 = 1.7$

17. From Figure 8a in Manual $= E_{mu}/E_{og} = 1.6$

18. $E_{mu} = 1.10$

19. $S' = t_p - Z_f$
    where t_p = plate spacing = 20 inch
    and $S' = 20.0 - 37.6 = 16.24$ inch

20. $U_g/S' = 2.03/16.24 = 0.125$

21. From Figure 11 in Manual $\varepsilon^o \sigma = 0.22$
    where σ = surface tension in dynes = 31.2 dyne/cm
    and $\varepsilon^o = 0.22/31.2 = 0.0074$

22. $r_e = 449 \times 0.0014\, V\rho_g/Q\rho_L = 449 \times 0.0074 \times 3488.1/3645.4 = 3.18$

23.
$$E_a = \frac{E_{mu}}{1+r_e E_{mu}} = \frac{1.10}{1.0+(3.18 \times 1.10)} = 0.25$$

24.
$$E_o = \frac{\log[1.0+E_a(\lambda-1)]}{\log \lambda}$$
$$= \frac{\log[1+0.25(2.39-1.0)]}{\log 2.39}$$
$$= 0.34 \text{ or } 34.0\%$$

8.2.5 Confirmation of plate efficiency

The efficiency estimated above applies strictly to a bubble cap column and has been adapted above for the sieve tray proposed and therefore this estimate must be checked. In addition, as also stated above, the F-factor was slightly less than the lower limit. Consequently the above estimate has been checked by applying O'Connell's chart [41]. From Figure 8.1 the relative volatility at the bottom of the column $\alpha = 2.64$. The viscosity of the feed at its boiling point is $0.62/cP$. The abscissa for O'Connell's chart $= 1.64$ and the overall tray efficiency $= 0.42$. Since both methods of predicting the tray efficiency are approximate but both agree reasonably well the mean will be taken. That is, a tray efficiency of 0.38 will be accepted for this design.

8.2.6 Evaluation of column height

In Section 8.2.3 and in Figure 8.1 it was shown that 13 ideal trays were required for the separation, so that for a tray efficiency of 0.38 the total number of actual plates required is 35 plates and the feed plate will be plate number 24 from the bottom plate of the column.

The dimensions of the column will be:

(i) Total height of all plates	= 17.8 m.
(ii) Height for connection to reboiler	= 1.0 m.
(iii) Vapour space height at top of column	= 0.2 m.
Total height of column	= 19.0 m.
(iv) Diameter of column	= 1.0 m.
(v) Sieve plates containing 4400 holes arranged on a triangular pitch (1.27 cm)	= 0.0127 m.
(vi) Hole size in sieve plate	= 0.47 cm.
(vii) Downcomer, segmental with a weir length	= 0.7 m.

8.3 Feed heater of product distillation column

The feed to the product distillation column is available from the intermediate storage tank at 35°C and will be fed to the column at the 1514.98 kg/h. From the temperature-equilibrium diagram [42] the bubble point of the feed will be 82°C, and the heat load of the feed heater will be

Heat required by MEK/alcohol:

MEK: 1345.60 × 2.2986 × (82.0–35.0)   = 1.4537 × 10⁵ kJ/h
Alcohol: 169.24 × 2.8052 × (82.0–35.0)   = 0.2231 × 10⁵ kJ/h
TCE: 0.14 × 1.1137 × (82.0–35.0)   = 0.0001 × 10⁵ kJ/h
                Total   = 1.6769 × 10⁵ kJ/h

The volumetric flowrate of the feed is:

$$\left(\frac{169.24}{807.0} + \frac{1345.60}{805.0} + \frac{0.1400}{1325.0}\right) = 1.8814 \text{ m}^3/\text{h}$$

For a heat exchanger the recommended velocity through the tubes should be 1.0 m/s. On this basis the sectional area will be

$$1.8814/3600 = 0.0005 \text{ m}^2 \equiv 5.226 \text{ cm}^2$$

or a single tube 2.58 cm in diameter. This implies that a concentric pipe heat exchanger is required with the feed passing through the central tube and steam condensing in the annulus. Choosing 2.54 cm nominal bore tube for the central tube and a 5.25 cm nominal bore pipe for the annulus the length of heat exchanger required may be estimated as follows.

8.3.1 Tube side coefficient of heat transfer, h_i

Bore of inner tube = 0.0234 m
Flow area = 4.29 × 10⁻⁴ m²
Velocity of feed through the tube = 1.8814/4.29 × 10⁻⁴ × 3600 = 1.218 m/s.
Reynolds number = (0.023 × 804.0 × 1.218)/(0.4 × 10⁻³) = 56308
From Kern (Figure 24), j_H = 160

$$h_i = \frac{160k}{d} Pr^{0.33} \left(\frac{\mu}{\mu_w}\right)^{0.14}$$

where k is thermal conductivity of the feed.

The thermal conductivity of MEK has not been reported whereas Kern[20] gives a value for k_{alc} = 0.1575 W/mK. However Perry[19] presents a method of predicting the thermal conductivity of liquids. In various units:

$$k = 1.034 C_p \, \rho^{4/3}/\alpha M^{0.33} \text{ Btu/h ft °F} \qquad (8.9)$$

where

$$\alpha = \frac{L_{vb}/T_b}{21.0} \qquad (8.10)$$

For 2-butanol $C_p = 0.66$, $\rho = 0.807$ g/ml, $L_{vb} = 9931.08$ cal/g–mole, $T_D = 371$ K.

Then
$$\alpha = 9932.08/(21.0 \times 371) = 1.275$$

and
$$k_{alc} = \frac{1.034 \times 0.66 \times 0.807^{1.33}}{1.275 \times 74.12^{0.33}} = 0.097 \text{ Btu/h ft }°\text{F}$$
$$= 0.1682 \text{ W/m K}$$

The agreement is very good and therefore the method will be applied to MEK. The pertinent properties are: $C_p = 0.55$, $\rho = 0.804$ g/ml, $L_{vb} = 7642.6$ cal/g–mole, $T_b = 352$ K.

Then
$$k_{MEK} = \frac{1.034 \times 0.55 \times 0.804^{1.33} \times 352 \times 21.0}{7642.6 \times 72.1^{0.33}}$$
$$= 0.0992 \text{ Btu/h ft }°\text{F} = 0.1717 \text{ W/m K}$$

and the thermal conductivity of the feed is 0.1713 W/m K.

The Prandtl number is
$$2.295 \times 10^3 \times 0.3 \times 10^{-3}/0.1713 = 4.02$$

Then
$$h_i = 160 \times 0.1713 \times 4.02^{0.33}/0.0234 = 1862 \text{ W/m}^2 \text{ K}$$

8.3.2 Annulus coefficient of heat transfer h_o

It is proposed that the feed be heated by condensing steam. Dry saturated steam is available at 140°C and this will be utilised, and to prevent the accumulation of condensate the heater will be arranged vertically. Then the equivalent diameter is (5.25−2.34) = 2.91 cm and the quantity of steam required is:-

$$(1.6769 \times 10^5)/(2.145 \times 10^3) = 78.2 \text{ kg/h}$$

where the latent heat of steam at 140°C (3.45 bar) = 2.145×10^3 kJ/kg,

The loading on the outside of the tube wall will be

$$G' = 78.2/(\pi \times 3.175 \times 10^{-3}) = 7839.95 \text{ kg/h m}$$

and the Reynolds number is

$$\frac{4G'}{\mu} = \frac{4 \times 7839.95}{0.014 \times 10^{-3} \times 3600} = 6.22 \times 10^5$$

The mean temperature of the condensing film will be assumed to be 100°C on the basis that the heat transfer coefficient of the condensing steam is 1900 W/m² K. The average heat transfer coefficient for the condensing surface becomes from Kern[20] page 265,

$$\bar{h}_o = 0.9247 \left[\frac{4k_f^3 \rho_f^2 g}{\mu_f^2} \cdot \frac{\mu_f}{4G'} \right]^{1/3} \tag{8.11}$$

$$= 0.9427 \left[\frac{4.0 \times 0.723^3 \times 1000^2 \times 9.807}{0.25 \times 10^{-3} \times 4.0 \times 2.178} \right]^{0.33}$$

$$= 1894 \text{ W/m}^2 \text{ K}$$

8.3.3 Heat transfer coefficient of metal tube

Kern[20] states that the resistance to heat transfer of the tube may be significant and should be included in the calculations. This will be included in the calculation of the overall heat transfer coefficient.

8.3.4 Overall heat transfer coefficient

The overall heat transfer coefficient will be based on the outside tube diameter and a scale resistance of 0.005 m²/k W will be included. Thus:-

$$\frac{1}{U} = \frac{1}{1894} + \frac{3.18}{2.34} \left(\frac{1}{1864} \right) + \frac{3.18}{2.76} \left(\frac{0.0042}{45.0} \right) + 0.0005$$

$$= 0.0005 + 0.0007 + 0.0001 + 0.005$$

$$= 0.0019$$

and $U = 536.0 \text{ W/m}^2 \text{ K}$

8.3.6 Heat transfer area

Table 8.2 – Data for log mean temperature difference (Feed heater).

Hot Fluid	Feed heater	Cold Fluid	Difference
140.0	Higher temperature	82.0	58.0
140.0	Lower temperature	35.0	105.0
0.0	Difference	47.0	47.0

From Table 8.2 the log mean temperature difference = 79.2°C

The heat transfer area required to heat the feed to the product distillation unit is:

$$\frac{1.6769 \times 10^5}{536 \times 79.2 \times 3600} = 1.10 \text{ m}^2$$

The heat transfer area per metre length of tube is 0.100 m² and therefore the length of the feed heater will be 11.0 m. Since the feed plate of the product distillation column is to be 13.2 m above the base of the column it will be convenient to arrange for the feed heater to be located vertically alongside the column in a single run of 11.0 m. Steam should be admitted into the annulus near the top adjacent to the feed entry point and provision should be made for steam condensate to be removed through a suitable trap so that no condensate accumulates in the annulus of the heater.

This concludes the design of the product distillation column feed heater.

8.4 Product distillation column condenser

The condenser of this distillation unit must be designed to condense 3488.1 kg/h of vapour leaving the top of the column at a temperature of 80.0°C. This vapour will be condensed in the shell of a shell and tube heat exchanger arranged horizontally. Cooling water will be circuited through the tubes and will enter the condenser at 24°C.

The heat load on the condenser is:—

MEK: 3453.22 × 443.51   = 1531537.60 kJ/h

Alcohol: 34.88 × 562.61   = 19624.26 kJ/h

TOTAL   = 1.5512×10^6 kJ/h

8.4.1 Condenser cooling water requirement

Cooling water is available at 24°C and the maximum temperature of the effluent cooling water must not exceed 49°C. Therefore the minimum amount of cooling water required for this condenser is:

$$(1.5512 \times 10^6)/(4.1868 \times 25.0) = 14820.0 \text{ kg/h}$$

corresponding to a flowrate of 0.0041 m³/s. Taking the optimum water velocity through the tubes to 1.0 m/s the minimum flow area required for the cooling water is 0.0041 m². However, the heat load is large and a substantial heat transfer area may be required. Furthermore it is preferable that the condenser be squat so that the accumulation of any non condensable gases will have a minimal effect and can be easily vented. On this basis a four-pass water-side/single-pass vapour-side condenser would appear to be appropriate and a condenser 38.75 cm shell diameter containing 140 tubes 1.9 cm, o.d., number 16 b.w.g. arranged on a 2.38 cm triangular pitch with 35 tubes per pass is envisaged. For this condenser the cooling water flow area will be (from Table 10 of Kern) 35 × 0.00195 = 0.0068 m², and in order to prevent the deposition of silt, etc, the water velocity will be maintained at 1.0 m/s in this design. For this velocity the volumetric flowrate is 0.0068 m³/s ≡ 24480 kg/h and the rise in temperature of the water will be 15.1°C, and the exit water temperature would be 39.1°C if the cooling water were obtained directly from the water supply. However, as will be seen in Section 8.5. the heat load required to cool the MEK from its boiling point to 30°C is 156025.79 kJ/h. Therefore if the cooling water was first passed through the product cooler the rise in temperature over the cooler would be 156025.79/(24480. × 4.1868) = 1.5°C. The exit water temperature from the MEK product cooler would be 25.5°C and this water could be used to condense the vapours discharged from the distillation column. The temperature of the water discharged from the condenser would be 40.6°C; considerably below the maximum effluent temperature of cooling water.

The above method of utilising the cooling water is considered suitable and will be the basis for the design of this item.

8.4.2 Cooling water heat transfer coefficient

The cooling water is to be circuited through the tubes at a water velocity of 1.0m/s and, from Kern[20] (Figure 25) the water side heat transfer coefficient will be $h_i = 3407$ W/m² K

8.4.3 Shell-side heat transfer coefficient

The vapour passing from the top of the column will enter the shell side of the condenser at 80.0°C and, following standard practice, segmental baffles will be included to ensure flow across the tubes. These will be arranged at half the maximum spacing; that is 20.0 cm baffle spacing will be introduced when the crossflow area is:

$$a_s = \frac{D \times C' \times B}{1000 P_T} = \frac{38.75 \times 0.475 \times 20}{2.38 \times 1000} = 0.016 \text{ m}^2$$

As a first trial let the tube length be 1.5 m and the loading is:—

$$G'' = 3488.1/(1.5 \times 140^{2/3}) = 86.23 \text{ kg/h}$$

The Condensate Reynolds number is:

$$(4 \times 86.23)/(0.23 \times 10^{-3} \times 3600) = 417.$$

and

$$\bar{h}_o = \left(\frac{(0.23 \times 10^{-3})^2}{0.172^3 \times 804^2 \times 9.807}\right)^{1/3} = \left(\frac{1.5}{417^{0.33}}\right) = 1701 \text{ W/m}^2 \text{ K}$$

8.4.4 The overall heat transfer coefficient

The overall heat transfer coefficient will be based on the outside tube diameter and, following the recommendation of Kern will include the tube wall resistance and a scale resistance of 0.0005. Then

$$\frac{1}{U_o} = \frac{1}{1701} + \frac{0.019}{0.016}\left(\frac{1}{3407}\right) + \frac{0.00165 \times 0.0174}{45.0 \times 0.016} + 0.0005$$

$$= 0.0006 + 0.0003 + 0.00004 + 0.0005$$

$$U = 700 \text{ W/m}^2 \text{ K}$$

8.4.5 Heat transfer area and condenser length,

The condenser length depends on the overall heat transfer coefficient and the temperature difference (Table 8.3).

Table 8.3 — *Data for log mean temperature difference (Condenser).*

Hot Fluid	Product still condenser	Cold Fluid	Difference
80.0	Higher temperature	40.6	39.4
80.0	Lower temperature	25.5	54.5
0.0	Difference	15.1	15.1

The log mean temperature difference = 46.6°C and the heat transfer area

$$= \frac{1.5512 \times 10^6 \times 10^3}{700 \times 46.6 \times 3600} = 13.2 \text{ m}^2$$

The heat transfer surface available per linear metre of condenser = 8.38 m²/metre length. Required length of condenser = 13.2/8.38 = 1.57 m. For safety a condenser tube length of 1.60 metres will be recommended. With headers for the inlet water and outlet of 0.25 m each the total length of the condenser will be 2.10 m.

The pressure drop calculations for a condenser were presented in Section 5.5 and will not be repeated here. Hence this concludes the design of the condenser.

8.5 Product cooler

The MEK product cooler is required to cool 1357.61 kg/h of distillate from its boiling point 80°C to 30°C. The heat load for this duty is

$$1357.61 \times 2.2986 \times (80 - 30) = 156025.79 \text{ kJ/h}$$

In Section 8.4.1 it was decided to use the same cooling water for this cooler and for the condenser of the distillation column by passing the water through this cooler before it passes through the condenser. The water rate through the cooler is therefore fixed at 24480 kg/h and the temperature rise through the cooler will be 1.5°C.

The calculation of the overall heat transfer coefficient is similar to that of the tube side of the feed heater and will not be repeated. The results of the calculations are that a cooler 20.0 cm o.d. shell diameter containing 36 tubes 1.9 cm o.d. number 16 b.w.g. arranged on a triangular pitch 2.38 cm with single pass shell side and single pass tube side would be suitable. The water would be passed through the tubes. The overall heat transfer coefficient is taken to be 625 W/m² K, and the log mean temperature difference is 22.0°C. On the basis of the above heat load a heat transfer area of 3.15 m² is required and since the surface area per linear metre for this heat exchanger is 2.15 m², the length of the tubes will be 1.5 metres. Thus a suitable heat exchanger will be one of 0.2 m shell diameter of overall length 2.0 m.

8.6 Product distillation column reboiler

The reboiler temperature will be 100°C and it will be required to generate 3488.1 kg/h of vapour containing 99.0% weight of 2-butanol. The heat load on this reboiler will be

$$3453.57 \times 562.62 = 1943047.5 \text{ kJ/h}$$
$$34.53 \times 443.51 = 15314.3 \text{ kJ/h}$$
$$\text{TOTAL} = 1958361.8 \text{ kJ/h}$$

The detailed design of a reboiler was presented in Section 4.10.2 and the calculations will not be repeated here. The allowable heat flux for a thermo-syphon natural circulation reboiler is 37800 W/m and therefore the heat transfer area required would be 1.0 m². This reboiler will be steam heated since dry saturated steam is available at 140°C. The temperature difference is adequate but much less than the critical temperature difference.

This concludes the design of the product distillation unit.

Chapter 9

THE HEAT BALANCE

Heat balances have been prepared for the major plant items and the detailed calculations given in particular sections of the previous chapters. In each balance the total heat into and out of the particular unit has been evaluated with respect to a reference temperature of 273 K. Since the detailed calculations associated with each item have been presented in the particular chapter describing the design of the unit all the heat balances have been presented in flow sheet form in Figures 9.1 to 9.4. These are self explanatory when considered with the detail design calculations.

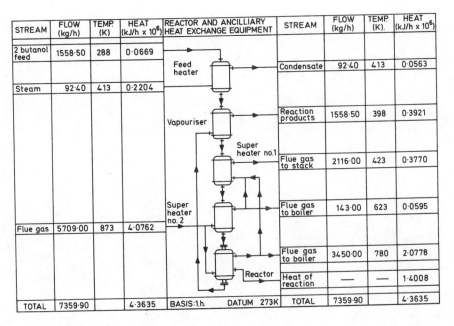

Figure 9.1 — *Heat flow diagram for the reactor and its ancillary heat exchange equipment.*

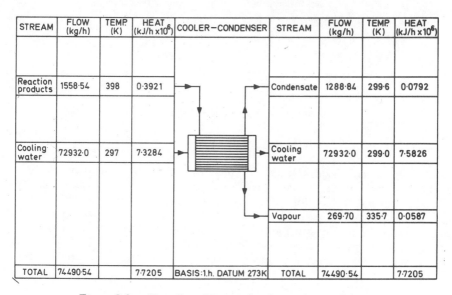

Figure 9.2 — *Heat flow diagram for the cooler-condenser*

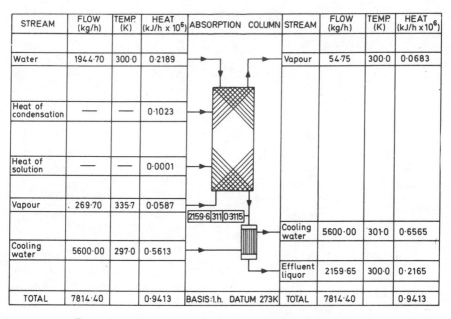

Figure 9.3 — *Heat flow diagram for the absorption column.*

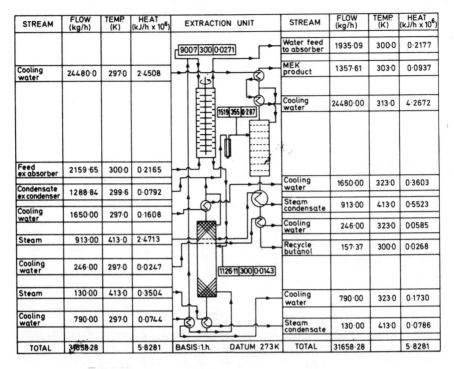

Figure 9.4 — *Heat flow diagram for the extraction unit.*

Figure 9.4 includes heat balances on the solvent recovery still and the product distillation unit because these items of plant are inter-related through the different recycle streams.

All heat flow diagrams have been based on one hour's throughput and the heat balances are calculated relative to a datum of 273 K.

137

Chapter 10

MECHANICAL DESIGN OF REACTOR

10.1 General discussion

The maximum operating temperature is 800 K (527°C) and therefore the design temperature will be taken as 550°C and maximum operating pressure is 2.5 bar (Section 4.8) and for safety the design pressure will be taken as 10 bar on the tube side and shell side to account for the possibility of blockage of the reactor tubes. Throughout the mechanical design all the items were designed to the specifications of British Standard BS5500: 1976[43].

10.2 Materials selection

The reactor is operating under fairly low pressure but at a temperature of 550°C and therefore the tube material selected must be resistant to hydrogen, produced by the reaction, and also to sulphur oxides which may be present in the flue gas, employed for heating. Nelson[44] has reported the effects of hydrogen attack by decarburisation and intergranular cracking on several types of ferrous alloy. He concluded that low alloy steels are completely unsuitable but that austenitic stainless steel is resistant to attack under the prevailing conditions providing that the metal has a low phosphide and sulphide content. Therefore it was decided to use stainless steel type 304S59 (which contains only 0.3% sulphur and 0.04% phosphorous) in the manufacture of the tubes, tubesheet and shell of the reactor. The most important consideration in selection of the tube material is the creep resistance at the elevated operating temperature and assuming that the reactors are designed to operate for approximately 15 years then the rupture stress[45] for steel type 304S59 should not exceed 108 N/mm^2. The design strength is therefore recommended as 81 N/mm^2 for a design lifetime[45] of 150 000 h and also the tube should be inspected after 100 000h operation. Seamless tubes are necessary since welded tubes are unsuitable at elevated temperatures.

10.3 Tube thickness

This has previously been assumed in heat transfer calculations (Section 4.7) as 3.68 mm but it is necessary that the tube thickness be adequate to withstand the design pressure at the design temperature. For cylindrical shells subject to internal pressure, the thickness required is given by

$$t = \frac{pD_o}{2f - p} + c \qquad (10.1)$$

where t = thickness required
 p = design pressure = 10 bar
 D_o = outside diameter of tube = 48.3 mm

f = design stress = 81 N/mm² for stainless steel type 304S59 tube at the design temperature [4,5]

c = corrosion allowance = 2 mm

t = (10 × 0.10133 × 48.3)/{(2 × 81) − (10 × 0.10133)] + 2.0 mm
   = 0.304 + 2.0 mm = 2.304 mm

Tube Thickness of 3.68 mm will be adequate

10.4 Design of tube sheets

The reactor is to be designed as a single pass shell and tube exchanger with fixed tube plates. The minimum tube plate thickness less a corrosion allowance for 48.3 mm o.d. tubes is recommended to be 30 mm (this value will be assumed in the trial and error design of the tube sheets). Since inspection of the tubes is necessary and to facilitate cleaning of the shell side by removal of the tube bundle, the tube sheets will be gasketed and bolted to the shell flanges. Therefore the thickness of the un-tubed annular ring may be determined by considering the design of a flat plate.

For a circular head, $t = CD\sqrt{(P/f)}$ where D = diameter of tube plate which is assumed to be the shell i.d. initially, and $C = 0.41$.

Therefore thickness of annular ring $t = 0.41 \times 736.6\sqrt{(1.0133/81)}$ mm = 33.8 mm

The tubesheet will be designed on the pressure existing on the tubeside since this is the larger.

10.4.1 *Effective tube side design pressure*

$$\text{Mean Strain ratio, } K = \frac{E_s t_s (D - t_s)}{E_t t_t N (d - t_t)}$$

where D = outside diameter of shell = 755.6 mm
       d = outside diameter of tubes = 48.3 mm
       N = number of tubes in tubesheet = 100
       t_t = tube thickness = 3.68 mm
       E_t = elastic modulus of tube material
       E_s = elastic modulus of shell material = E_t

The shell thickness, t_s was calculated using equation (10.1) for a design pressure of 10 bar (to allow for tube failure) hence t_s = 6.7 mm. The standard wall thickness for this diameter pipe[22] is 9.52 mm and this value will be used for the shell thickness. t_s = 9.52 mm

$$\therefore K = \frac{9.52(755.6 - 9.52)}{3.68 \times 100 \times (48.3 - 3.68)} = 0.433$$

Tubesheet factor, $Fq = 0.25 + (F - 0.6)\left[\dfrac{300 t_s E_s}{KLE}\left(\dfrac{D_1}{t}\right)^3\right]^{0.25}$

where F = 1.0 for a gasketed and bolted tubesheet
E = elastic modulus of tubesheet material = E_s
D_1 = diameter to which shell fluid pressure is exerted = 736.6 mm
t = tubesheet thickness assumed to be 30 mm
L = tube length between tubesheets = 3.0 m

$$\therefore Fq = 0.25 + 0.4\left[\dfrac{300 \times 9.52}{0.433 \times 3000}\left(\dfrac{736.6}{30}\right)^3\right]^{0.25} = 5.623$$

$$f_t = 1 - N\left(\dfrac{d - 2t_t}{D_2}\right)^2$$

where D_2 = diameter to which tube fluid pressure is exerted = 736.6 mm

$$\therefore f_t = 1 - 100\left[\dfrac{48.3 - (2 \times 3.68)}{736.6}\right]^2 = 0.691$$

$$p'_t = p_2\left[\dfrac{1 + 0.4 J K(1.5 + f_t)}{(1 + JKFq)}\right]$$

where p_2 = the tube side design pressure = 1.0133 N/mm²
J = the expansion joint strain factor = 1.0 for a shell without expansion joint

$$\therefore p'_t = 1.0133\left[\dfrac{1 + 0.4 \times 0.433(1.5 + 0.691)}{(1 + 0.433 \times 5.623)}\right] = 0.407 \text{ N/mm}^2$$

$$p'_s = p_1\left\{\dfrac{0.4J[1.5 + K(1.5 + f_s)] - \tfrac{1}{2}(1 - J)(D_j^2/D_1^2 - 1)}{1 + JKFq}\right\}$$

where p_1 is the shell side design pressure = 1.0133 N/mm²
and $f_s = 1 - N(d/D_1)^2 = 1 - 100(48.3/736.6)^2 = 0.570$.

Therefore for $J = 1$

$$p'_s = 1.0133\left\{\dfrac{0.4[1.5 + 0.433(1.5 + 0.570)]}{1 + 0.433 \times 5.623}\right\} \text{N/mm}^2 = 0.283 \text{ N/mm}^2$$

Pressure due to differential thermal expansion, p_e is given by,

$$p_e = \dfrac{4JE_s t_s(\alpha_s \theta_s - \alpha_t \theta_t)}{(D - 3t_s)(1 + JKFq)}$$

where

θ_s is mean shell temperature less 10 K = 770 K
θ_t is mean tube temperature less 10 K = 740 K
$\alpha_s = \alpha_t$ is the thermal expansion coefficient [19] of stainless steel type 304S59
= 18.3 × 10^{-6} mm/mm K for expansion from 20 to 550° C
E_s = elastic modulus = 0.193 × 10^6 N/mm^2 [19].

$$\therefore p_e = \frac{4 \times 0.193 \times 10^6 \times 9.52 \times 18.3 \times 10^{-6}(770-740)}{(755.6 - 3 \times 9.52)(1 + 0.433 \times 5.623)} \text{ N/mm}^2$$

$$= 1.615 \text{ N/mm}^2$$

For a fixed tubesheet when extended for bolting to heads with ring type gaskets equivalent bolting pressure, $P_{Bt} = (2\pi/F^2)(M_1/D_1^3)$ where M_1 = the total moment acting upon the extension under operating conditions and is given by,

$$M_1 = H_D h_D + H_T h_T + H_G h_G$$

For integral type ring flanges (Class 150) welded to the shell side of inside diameter of 29 ins.

    flange thickness [46]   = 22.25 mm
    o.d. of flanges   = 879.5 mm
    bolt circle diameter, C   = 838.2 mm
    bolts are 24 of $\frac{3}{4}$.inch nominal size.

Using corrugated stainless steel gaskets filled with asbestos; gasket factor, $m = 3.50$. $N = 10$ mm, hence effective gasket width, $b_o = N/2 = 5$ mm. Therefore diameter at location of gasket load reaction is given by the mean diameter of gasket contact face. $G = 762$ mm i.d. of flange, $B = 736.6$ mm

$$\text{Moment arms, } h_D = \tfrac{1}{2}(C-B) = \tfrac{1}{2}(838.2 - 736.6) = 50.8 \text{ mm}$$

$$h_G = \tfrac{1}{2}(C-G) = \tfrac{1}{2}(838.2 - 762) = 38.1 \text{ mm}$$

$$h_T = \tfrac{1}{2}(h_D + h_G) = \tfrac{1}{2}(50.8 + 38.1) = 44.45 \text{ mm}$$

Hydrostatic end force, on inside of flange,

$$H_D = 0.785 \, B^2 p = 0.785 \, (736.6)^2 \, 1.0133 \text{ N} = 0.4316 \times 10^6 \text{ N}$$

Hydrostatic end force due to pressure on flange face,

$$H_T = H - H_D, \text{ where } H = 0.785 \, G^2 p$$
$$= 0.785 \, (762)^2 \, 1.0133 - (0.4316 \times 10^6) \text{ N}$$
$$= 0.0303 \times 10^6 \text{ N}$$

Gasket Load, $H_G = 2b_o \times 3.14 \, Gmp$
$$= 2 \times 5.0 \times 3.14 \times 762 \times 3.5 \times 1.0133 \text{ N}$$
$$= 0.0849 \times 10^6 \text{ N}$$

$$= (0.4316 \times 50.8) + (0.0303 \times 44.45) + (0.0849 \times 38.1 \times 10^6) \text{ Nmm}$$

Therefore total moment, $M_1 = 26.51 \times 10^6$ N mm

Equivalent Bolting Pressure, $P_{Bt} = (2\pi/F^2)(M_1/D_1^3)$, where $F = 1.0$ for gasketed tube sheet

$$= 2\pi \times 26.51 \times 10^6 / 736.6^3 \text{ N/mm}^2$$

$$= 0.4167 \text{ N/mm}^2$$

Based upon bending stresses, effective tube side design pressure

$$p_2' = \tfrac{1}{2}(p_t' + p_{Bt} + p_e) = \tfrac{1}{2}(0.407 + 0.4167 + 1.615)$$

So for bending stresses, $P_2' = 1.219$ N/mm²

For shear stresses, $p_2' = \tfrac{1}{2}(p_t + p_e) = \tfrac{1}{2}(0.407 + 1.615) = 1.011$ N/mm²

10.4.2 Tube sheet thickness

For bending stresses

$$t = \tfrac{1}{2}FD_2\sqrt{(P_2'/f)}$$
$$= \tfrac{1}{2}(1.0 \times 736.6)\sqrt{(1.219/81)} \text{ mm}$$
$$= 45.2 \text{ mm}$$

For sheer stresses,

$$t = 0.155 D_0 p_2' / \lambda \tau$$

where λ = ligament efficiency = $[P_T - (d_h - t_t)]/P_T$, P_T being the tube pitch = 60.3 mm, so that $\lambda = [60.3 - (48.7 - 3.68)]/603 = 0.248$.

τ = shear design stress = $0.5f = 40.5$ N/mm²

$$\therefore t = \frac{0.155 \times 676 \times 1.011}{0.248 \times 40.5} \text{ mm} = 10.5 \text{ mm}$$

Since the value calculated from bending stresses is considerably larger than this, the former will be accepted for the tubesheet thickness.

Tube sheet thickness = 45.2 mm

10.5 Shell thickness,

This has been shown to be 9.52 mm for the cylindrical section of the shell but it is necessary to evaluate the thickness of the domed ends assuming that tori-spherical ends are to be used. Using the procedure described by BS5500:1976 the following limitations on the design are imposed, thickness, t_{se} $0.002D_i \leqslant t_{se} \leqslant 0.08D_i$, i.e. $1.47 \leqslant t_{se} \leqslant 58.9$ mm

small radius, $r \geqslant 0.06D$ and $r \geqslant 2t_{se}$, i.e. $r \geqslant 44.2$ mm

large radius, $R \leqslant D$, i.e. $R \leqslant 736.6$ mm

Ratio of design pressure to design stress of material is given by,
$$p/f = 1.0133/81 = 0.0125$$
Height of curved section of domed end, h_e is given by $h_e = D^2/4(R + t_{se})$ then assuming initially that $R = 736.6$ mm, thickness, $t_{se} = 9.52$ mm,
$$h_e = 736.6^2/4(736.6 + 9.52) \text{ mm} = 181.8 \text{ mm}$$
then $h_e/D = 181.8/736.6 = 0.247$ from BS5500: 1976 for $p/f = 0.0125$ and $h_e/D = 0.247$ then $t_{se}/D = 0.0067$.

Therefore thickness, $t_{se} = 0.0067 \times 736.6 = 4.9$ mm + 2 mm corrosion allowance. Therefore a shell domed end thickness of 9.52 mm will be adequate for the design conditions.

10.6 Branch connections

10.6.1 *Diameter of connections*

Firstly, it is necessary to evaluate an economic pipe diameter for the shell and tube fluids. For the tube side, density of vapour = 1.308 kg/m³, and volumetric flowrate of vapour = 0.3309 m³/s. Using the nomogram given by Perry [19] for schedule 40 steel pipe, economic pipe diameter = 178 mm i.d. Since the nomogram was evaluated for mild steel pipe, the higher costs of the stainless steel pipe employed in the reactor necessitates division of the pipe diameter by a factor of say 1.25 then economic pipe diameter = 178/1.25 = 142 mm i.d. The nearest standard pipe diameter is 6 inch nominal size[22] of $d_i = 154.08$ mm, $d_o = 168.3$ mm.

10.6.2 *Reinforcement of branch connections*

Assuming that the connections are fixed flush to the internal diameter of the shell on the cylindrical sections. Let the collar thickness be 5 mm then extended shell thickness, $T_r = 14.52$ mm then

$$\rho = \frac{d_i}{D_0} \sqrt{\left(\frac{D_0}{2T_r}\right)}$$

$$= \frac{154.08}{755.6} \sqrt{\left(\frac{755.6}{2 \times 14.52}\right)}$$

$$= 1.040$$

For negligible external loads operating in the creep range, $C = 1.0$ then $CT_r/T = 1.525$. Also diameter ratio $d_i/D_o = 0.204$ then nozzle thickness.

$$t_{r1} = 0.425T_r = 6.171 \text{ mm}$$

$$t_{r2} = d_i Y \sqrt{\left(\frac{T_r}{2D_0}\right)}$$

$$= 154.08 \times 0.51 \sqrt{\left(\frac{14.52}{2 \times 755.6}\right)} \text{ mm}$$

$$= 7.703 \text{ mm}$$

interpolating, nozzle thickness

$$t_r = t_{r1} + 10[(d_i/D_0) - 0.2](t_{r2} - t_{r1})$$

$$= 6.171 + 10(0.204 - 0.2)(7.703 - 6.171) \text{ mm}$$

$$= 6.23 \text{ mm}$$

Since the tube wall thickness of the connections is 7.11 mm then no reinforcement of the nozzle is required. The extent of reinforcement, H (width of collar) was calculated to be 77.04 mm. The branch flanges were designed[47] to BS4504:1969 and the dimensions are summarised in Figure 10.1.

10.6.3 *General considerations*

The brass catalyst particles (right circular cylinders of 3.2 mm diameter and length) will be retained in the reaction tubes using a stainless steel mesh attached to both ends of the tubes. A plain weave type of mesh of 1.626 mm wire diameter, 2.607 mm aperture diameter and 38% open area will be adequate to support the catalyst particles with minimal pressure drop.

10.7 Design of ancillary equipment

10.7.1 *General discussion*

In this chapter the design of pumps, storage vessels and pipe sizes are presented. The designs of two pumps are made in detail, *i.e.* the alcohol feed pump and the vapour feed compressor which involve the handling of liquid and vapour respectively. As the procedure for design of the remaining pumping equipment is identical, only the power requirements have been specified.

10.7.2 *Design of alcohol feed pump*

It is-proposed to employ this feed pump to deliver cold feed from the feed storage vessel through the cold feed preheater, vertical thermo-syphon reboiler, knock-out drum to the exit of the second vapour preheater. At first, the head required to reach the reboiler is evaluated. All physical properties are calculated at 333 K, the mean liquid temperature.

Liquid flowrate $= 1558.5/(3600 \times 761.2) \text{ m}^3/\text{s} = 5.687 \times 10^{-4} \text{ m}^3/\text{s}$

For a linear velocity in the pipe of 1.5 m/s, pipe area $= 5.687 \times 10^{-4}/1.5 \text{ m}^2 = 3.791 \times 10^{-4} \text{ m}^2 = \pi d^2/4 \text{ m}^2$, so pipe diameter $= 0.0219$ m. Therefore a ¾ inch nominal size pipe will suffice of $d_i = 0.02096$ m.

Taking a typical pipe roughness of 0.00018 the relative roughness, $e/d_i = 0.00018/0.02096 = 0.0087$.

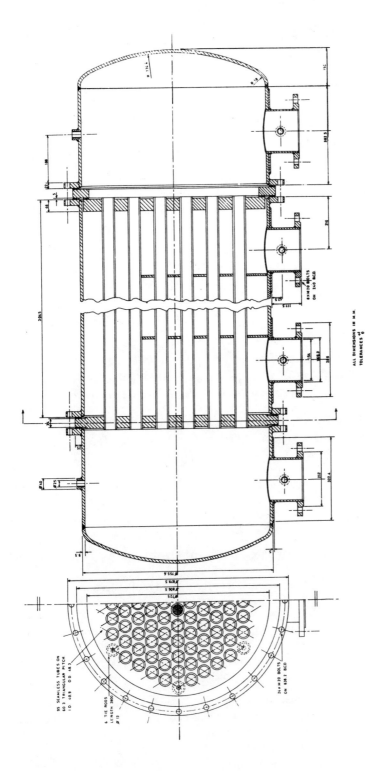

Figure 10.1 – *Tubular catalytic reactor for dehydrogenation of 2-butanol to 2-butanone (MEK)*. Design temperature 823K. Design pressure 10 bar. Material of construction stainless steel type 304S59. Designed to specifications of BS 5500; 1976, BS 3274: 1960, BS 4504: 1969.

viscosity, $\mu = 0.240 \times 10^{-3}$ kg/m s (Appendix H)
density, $\rho = 761.2$ kg/m³ (Appendix H)
velocity, $v = 1.65$ m/s

$$\therefore Re = \frac{dv\rho}{\mu} = \frac{0.02096 \times 1.65 \times 761.2}{0.240 \times 10^{-3}} = 109700$$

hence from chart presented by Coulson and Richardson[10] $R/\rho v^2 = 0.0045$

The equivalent length of pipe must now be estimated bearing in mind that three valves are required in this section. For valves, assume 200 pipe diameters and for other miscellaneous fittings, assume a total of 300 pipe diameters. Also, the length of pipe will not be known but is assumed to be 60m. Therefore equivalent length of pipe, $L_e = 60 + 02096\,[(3 \times 200) + 300]$m $= 79$m.

$$\therefore \text{Head due to friction}^{10}, h_f = 4\left(\frac{R}{\rho v^2}\right)\left(\frac{L_e}{d}\right)\left(\frac{v^2}{g}\right) \quad (10.2)$$

$$= \frac{4 \times 0.0045 \times 79 \times 1.65^2}{0.02096 \times 9.81}\,\text{m}$$

$$= 18.8\,\text{m}$$

In addition, the pressure drops across each unit must be considered except that of the reboiler which is compensated by the leg from the knock-out drum.

Pressure drop across feed preheater = 0.001 bar (Section 4.10.2)
Pressure drop across knock out drum = 0.05 bar (assumed)
Total = 0.05 bar

Therefore Head loss = 0.7 m, and total head to be developed = 18.8 + 0.07 m = 19.5m neglecting kinetic energy head.

$$\therefore \Delta P_f = 19.5 \times 761.2 \times 9.81\,\text{N/m}^2 = 145\,613\,\text{N/m}^2$$

$$\therefore \text{Power required} = 145613 \times 5.687 \times 10^{-4}/745\,\text{h.p.} = 0.11\,\text{h.p.}$$

Assuming that a centrifugal pump is selected for the duty, of efficiency 60%, then power required = 0.111/0.6 = 0.185 h.p.

$\therefore$ A 0.2 h.p. Centrifugal pump will perform the required duty

10.7.3 Design of vapour compressor

For this purpose, a single acting compressor will be used to transfer the 2-butanol vapour through the vapour superheaters, reactor, thermo-syphon reboiler and condenser in a 6 inch nominal size pipe.

Considering the heat losses due to friction which were calculated as in the preceding section with the physical properties of the vapour evaluated at 573 K

Vapour flowrate = 1558.5/(3600 × 1.574) = 0.275 m³/s
For a linear velocity of 15 m/s, pipe area = 0.275/15 m² = 0.0183 m²
= $\pi d^2/4$ m²
Pipe diameter = 0.153 m

Therefore 6 inch nominal size i.p.s. tube will suffice of d_i = 0.1541 m, and actual velocity = 14.7 m/s.

Relative roughness, e/d = 0.0012
viscosity, μ = 1.567 × 10^{-5} kg/m s (Appendix H)
density, ρ = 1.574 kg/m³ (Appendix H)
velocity, v = 14.7 m/s

$$\therefore Re = \frac{dv\rho}{\mu} = \frac{0.1541 \times 14.7 \times 1.574}{1.567 \times 10^{-5}} = 227540$$

hence from the chart[10], $R/\rho v^2$ = 0.00225.

In this section, assuming that there are 10 valves
∴ Equivalent length of pipe, L_e = 60 + 0.1541 [(10 × 200) + 300] m = 414 m
From equation (10.2)
∴ Head loss due to friction, h_f = 4 × 0.00225 × 414 × 14.7²/0.1541 × 9.81 m = 532 m
∴ ΔP_f = 532 × 9.81 × 1.574 N/m² ≑ 0.08 bar considering the pressure drops through the units

Pressure drop across superheaters	= 0.02 bar
Pressure drop through reactor	= 1.79 bar
Pressure drop through reboiler	= 0.59 bar
Pressure drop through condenser	= 0.02 bar
Pressure drop due to friction	= 0.08 bar
Total	= 2.50 bar

∴ pressure to be developed = 3.50 bar absolute
Mean Temperature of Vapour in section = 573 K
∴ Volumetric flowrate = 0.275 m³/s

Using a rotary compressor running at 1000 r.p.m. with a cylinder clearance of 4% and a stroke of 0.3 m, Volume per stroke = 0.275 × 60/1000 = 0.0165 m³
Compression ratio = 3.5/1.0 = 3.5

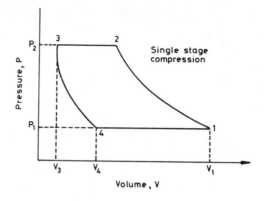

Figure 10.2 — *Vapour compressor cycle.*

Swept Volume, $V_s = (V_1 - V_3)$, Figure 10.2, and is defined thus

$$V_1 - V_4 = V_s[1 + c - c(P_2/P_1)^{1/\gamma}]$$

where $\gamma = 1.4$ if compression and re-expansion are assumed to be isentropic, and clearance, $c = 0.04$

$$\therefore 0.0165 = V_s [1 + 0.04 - 0.04(3.5)^{1/1.4}]$$
$$0.0165 = V_s [1.04 - 0.0979]$$

$\therefore V_s = 0.0165/0.942 = 0.0175$ m^3

$\therefore$ Cross sectional area of cylinder = $0.0175/0.3 = 0.058$ m^2

$\therefore$ Diameter of Cylinder = 0.272 m

Work of compression per cycle is given by

$$W = P_1(V_1 - V_4)\frac{\gamma}{\gamma - 1}\left[\left(\frac{P_2}{P_1}\right)^{(\gamma - 1)/\gamma} - 1\right]$$

$$= 1.0 \times 1.0133 \times 10^5 \times 0.0165 \left(\frac{1.4}{1.4 - 1}\right)(3.5^{0.4/1.4} - 1) \text{ J}$$

$$= 5852[1.43 - 1] \text{ J}$$

$$= 2519 \text{ J}$$

$\therefore$ Horse power required = $(2519 \times 1000)/(60 \times 745) = 56.4$ hp

Assume a compressor efficiency of 60%, then actual power = $56.4/0.6 = 93.9$ hp

$\therefore$ A 100 hp Rotary compressor will perform the required duty.

Chapter 11

ECONOMIC EVALUATION OF THE PROCESS

11.1 General discussion

This design study has been concerned with the production of MEK from 2-butanol for which the technology is not a limiting factor and design of the necessary equipment has been completed. However, in the final analysis, the proposed design can only be acceptable if the process is profitable. Also the choice of process route is based upon economic decisions when the alternatives are all technologically feasible. The decision to recover the MEK and 2-butanol from the vapour phase leaving the condenser was found to be economically more attractive than allowing all components in the vapour to be used as fuel, despite the necessity for increased capital investment. The comparison is presented at the end of this chapter, but initially an economic evaluation of the proposed process will be completed. This involves the acquisition of capital and operating cost data, and although published costs are used wherever possible, the lack of information which would be provided later by tenders from contractors, after compilation of final design details, necessitated the use of estimating techniques. If economically acceptable at this level of estimating more detailed design and costing exercises would be carried out to enable a more accurate profitability assessment to be carried out. This procedure might be repeated several times in practice until the final detailed estimate is produced.

11.2 Capital cost estimation

Available techniques of capital cost estimating include the ratio method, factor estimating and counting of "functional units" or other defined process steps. Since the accuracy of any cost estimate depends on the quantity or quality of information used in the preparation of that estimate, the detailed factor estimating approach can be the most accurate. For this method, experience factors (Lang factors) are used to determine the total capital cost using the delivered equipment cost and this approach has been adopted in the comparison of the alternative process routes described in Section 11.6. The ratio and step counting methods yield the total capital cost directly and they will be employed to find the total fixed investment for the proposed process plant.

11.2.1 *Ratio method*

Published capital costs for existing process plants may be adjusted on the basis of size, capacity or throughput in order to make an initial estimate, provided that the process routes are similar and that the capacities do not differ by more than five fold[48]. Haselbarth[49] reported that the capital cost of a plant producing 35 000 tons per year of MEK from 2-butanol was 3.75×10^6 in 1967. For the plant described in this study, the comparable capital cost would be given by an equation which represents the economy of scale[48]

$$C_1/C_2 = (Q_1/Q_2)^{0.6}$$

where C_1 is the capital cost of this designed plant
C_2 is the capital cost of the existing plant = \$3.75 × 10^6.
Q_1 = Throughput of plant 1 = 10800 tonnes/year
Q_2 = Throughput of plant 2 = 31818 tonnes/year
giving C_1 = \$3.75 (10800/31818)$^{0.6}$ × 10^6

Since the published value of capital cost is relevant to a plant built in USA, a location index will be employed for a plant built in the UK on an existing site. A value of 1.1 for this index was selected[50]. Also the value of C_2 is relevant to the year of 1967 and therefore must be updated to account for "inflation", and technological advance which leads to process design improvement. This is accomplished by means of a cost index thus:

Present cost = Previous cost × (Present index/Previous index)

Many cost indices exist, but the international EPE plant cost index[51] which is based on a value of 100 in January 1970 is recommended for use in the chemical industry and is available for 16 countries.

For 1967, EPE index for USA = 84.2
For June 1978, EPE index for USA = 180

Finally, it is necessary to convert from the dollar to sterling currency by dividing by 1.8, the current exchange rate, and the capital cost, C_1, is then given by,

$$C_1 = £3.75 \left(\frac{10800}{31818}\right)^{0.6} \left(\frac{1}{1.1}\right) \left(\frac{180}{84.2}\right) \left(\frac{1}{1.8}\right) \times 10^6$$
$$= £2.117 \times 10^6$$

11.2.2 *Functional unit approach*

A functional unit constitutes a significant step in a process and includes all equipment and ancillaries necessary for operation of that unit. Thus the sum of the costs of all functional units in a process gives the total capital cost. Generally a functional unit may be characterised as a unit operation, unit process, or separation method which involves energy transfer, moving parts or a high level of internals. All process streams are considered including side and recycle. Further definition of functional units and their use is described by Bridgwater [48] and for this process, seven functional units are proposed, namely; Compressor, Reactor, Absorber, Solvent extractor, Solvent recovery column, Main distillation column, Furnace and waste heat boiler.

The application of regression analysis to data from over 100 gas phase processes yielded the following expression for the cost of one functional unit on a battery limits basis[52]

$C_1/N = \$1.278 \times 10^{-3} (Q_1)^{0.616} \times 10^6$ for July 1976 where $N = 7$

Updating the cost and converting the currency as described in section 11.2.1.

$$C_1 = £7 \times 1.278 \times 10^{-3}(10800)^{0.616} \left(\frac{185}{152}\right) \left(\frac{1}{1.8}\right) \times 10^6$$

$$= £1.846 \times 10^6$$

11.2.3 "Process step scoring" method

This step counting method, which was developed by Taylor[53] is based on a system in which a complexity score accounting for factors such as throughput, corrosion problems and reaction time is estimated for each process step. From the list of "significant process steps" given by Taylor, the following were used to represent the steps for the MEK process; Feed Storage, Vaporiser, Reaction, Compress, Furnace and Waste Heat Boiler, Phase Separation, Absorber, Extractor, Solvent Recovery (Stripping), Product Distillation, Product Storage.

For each of these 11 process steps, a score is allocated which is summarised in Table 11.1. The total score for each step is related to "costliness" indices which are then added to give the "costliness" index, I, for the whole process. The battery limits capital cost is then determined using the following relationship[53]

$$C_1 = 42I \text{ (Capacity in 1000 tons)}^{0.39}$$

$$= 42 \times 17.8(10.607)^{0.39} \times 300/280 \text{ } £1000$$

$$= £2.012 \text{ million}$$

The accuracy of the above correlations are of the order of ±30% and therefore the results obtained for the battery limits capital cost are in good agreement. The mean of the values obtained using the three methods will be accepted as a reasonable estimate of the capital cost.

$$\text{Capital Cost, } C_1 = £1.992 \times 10^6$$

11.3 Operating costs

Variable or operating costs comprise all recurrent costs directly or indirectly incurred in manufacturing the product. It is common practice to express the total operating cost in terms of the raw material, direct labour, energy (or utilities) and fixed capital costs.

11.3.1 Raw material cost

This process for the production of MEK from 2-butanol has been assumed to be part of a large chemical plant complex and therefore the cost of the feedstock will be a "transfer" price taken as £7.77/100 kg. The supply of 2-butanol required is 1401 kg/h and the feedstock cost is given by

$$£7.77 \times 1401 \times 8000/100 = £871\ 206$$

Table 11.1 – Scoring for complexity of significant process steps.

Process Step	Through-put	Reaction Time	Storage Time	Minimum temperature	Maximum Temperature	Minimum Pressure	Maximum Pressure	Material Construct	Number of Streams	Total Score	Costliness Index
Feed Storage			1							1	1.3
Vaporise	1.5				0.5			1		3	2.2
Reaction					1			1		2	1.7
Compress										0	1.0
Furnace	3				1					4	2.8
Phase Separation										0	1.0
Absorber	3							1		4	2.8
Extractor	1							1		2	1.7
Stripping										0	1.0
Product Distillation										0	1.0
Product Storage			1							1	1.3
											17.8

For relationships between operating conditions and score and between total score and costliness index, see Taylor[53].

To this must be added the cost of trichlorethane make up. From the material balance, the solvent requirement is 1120 kg per annum and the cost[54] is £35.23/100 kg

$\therefore$ Cost of solvent = £394

Total Raw Material cost, R = £871 600

11.3.2 Direct labour cost

The direct operating labour costs for a chemical process may be expressed as[48] $L = £2074 \, N \, Q^{0.13}$ per year, where N is the number of functional units, Q is the plant capacity. This cost is relevant to 1976 and must be updated using a labour cost index. Therefore

$$L = £2074 \times 7 \times (10800)^{0.13} \left(\frac{300}{248}\right) = £58\,753$$

Table 11.2 — *Summary of plant utilities requirement.*

Equipment	Cooling Water (kg/h)	Equipment	Steam (kg/h)	Equipment	Electrical Power (kW)
Reaction product Condenser	72 932	Feed Heater	93	Compressor (100 hp)	74.57
Cooler for liquor from Absorber	5 600	Main Distillation Column Reboiler	913	Feed Pump (½hp)	0.37
Condenser for Solvent Stripper	1 650	Solvent Stripping Column Reboiler	130	7 Transfer pumps	2.61 (estimated)
Recycle solvent cooler	790			Total	77.55
Recycle 2-butanol cooler	246			Electrical power for ancilliaries say 10%	7.75
MEK product cooler	24 480				
Total	105 698	Total	1136	Total	85.30

11.3.3 Energy (utilities) cost

Since detailed energy balances have been completed in Chapter 9, sufficient data is available to calculate directly all utilities costs and the plant requirements are listed in Table 11.2. The furnace will be fired using heavy fuel oil (s.g.= 0.949) which produces 14.96 kg flue gas per kg of fuel[55] and for 8000 h operation per year

Volume of fuel oil required

$$= \frac{5709 \times 8000 \times 2.2046}{14.96 \times 9.49} \text{ gallons/year}$$

Cost of fuel oil at 30p per gallon = £212812. However, this cost is to some extent offset by the steam producing capacity of the flue gas which leaves the plant at 780 K. Heat content of flue gas used in boiler = 2.0778×10^6 kJ/h.

If gas is cooled to 423 K to produce steam at 3.6 bar, Heat content of gas to stack = 0.6147×10^6 kJ/h. The gross amount of heat available is 1.463×10^6 kJ/h. Assuming that the feed to the boiler is water at 293 K and steam is raised at a thermal efficiency of 80%, taking the specific heat of water at 353 K as 4.198 kJ/kg

Mass of steam produced,

$$W_s = \frac{1.463 \times 10^6 \times 0.8}{[4.198(413-293)+2145]} \text{ kg/h} = 441 \text{ kg/h}$$

Taking the cost of steam raising to be £8/1000 kg, cost of steam production = $(1136-441) \times 8000 \times 8/1000$ £/year = £44 480 per year.

Cost of cooling water at 3p per m³ is given by £ $105\,698 \times 8000 \times 0.03/1000$ per year = £25 367 per year

Cost of electricity at 2p per kWh = £85.30 $\times$ 8000 $\times$ 0.02 per year = £13 648 per year

Then the total utilities cost is the sum

Fuel oil	£	212 812
Steam	£	44 480
Cooling Water	£	25 367
Electricity	£	13 648
Total Energy (Utilities) Cost E	£	296 307

11.3.4 *Total Operating Charges*

In addition to the three principal direct costs as calculated above, the remaining constituent elements of the operating costs including fixed costs (rates, rent, property taxes, *etc*) indirect costs (safety, general overheads, packaging, research and development, inspection, *etc,*) and administration and distribution costs may be expressed as a function of raw material, direct labour, energy and the fixed investment costs. Evaluation of the total operating cost equation is shown in Table 11.3 using the method described by Bridgwater[4,9]

$$O = 1.05R + 1.9565L + 1.05E + 0.119C_1 + 0.0408[1.3(O)]$$
$$\therefore O = 1.109R + 2.066L + 1.109E + 0.126C_1 \qquad (11.1)$$

The annual operating cost may be determined by substitution of the direct and capital costs into equation (11.1) Thus,

$$O = (1.109 \times 871600)+(2.066 \times 58753)+(1.109 \times 296307)$$
$$+(0.126 \times 1992000) \quad \text{£/year}$$

$$= £1.6675 \times 10^6 \text{ per year excluding depreciation and interest}$$

Table 11.3 — Operating cost estimation.

	Range	R	L	E	C1
Direct costs					
a. Raw Materials		1.0			
b. Energy				1.0	
c. Labour			1.0		
d. Supervision	$0.1 L - 0.25 L$		0.17		
e. Payroll charges	$0.15 - 0.5 (c+d)$		0.585		
f. Maintenance	$0.02 C_1 - 0.15 C_1$				0.05
g. Operating supplies	$0.005 C_1 - 0.01 C_1$				0.005
h. Laboratory	$0.03 L - 0.2 L$		0.05		
j. Royalty	$0 - 0.6 S^*$				
Sub total		1.0	1.805	1.0	0.055
k. Contingency	$0.01 - 0.1 \, (a \to j)$	0.05	0.0903	0.05	0.0028
Indirect costs					
l. Rates	$0.02 C_1 - 0.04 C_1$				0.03
m. Insurance	$0.004 C_1 - 0.02 C_1$				0.01
n. Administration	$0.4 L - 0.8 L +$		0.06		0.02
p. Research	$0.015 S^* - 0.055 S^*$				
q. Distribution and Selling	$0.02 S^* - 0.2 S^*$				$(0.04 S)$
Sub total			0.06		0.06
r. Contingency	$0.01 - 0.05 \, (l \to n)$		0.0012		0.0012
Total		$1.05 R$	$1.9565 L$	$1.05 E$	$0.119 C_1$

*S is the selling price of the product which may be approximated as $1.3 \, (O)$ where O is the total operating cost.

11.3.5 Taxes and depreciation

Since the assessment of tax liabilities is financially complex for a project of this scale it will be assumed that a corporation tax of 52% be imposed on gross profit necessarily payable in the year following the time of imposition. Also a linear rate of depreciation of capital equipment is assumed over an operating life of 10 years with a sum of 5% of the initial investment being recoverable as scrap value at the end of this period.

11.4 Process evaluation

The cash flows for the proposed process design have been evaluated and it is essential to ensure that an adequate return will be obtained from the capital employed. There are a great many techniques for assessing profitability and three of the more useful will be applied to this process.

Table 11.4 — Evaluation of net cash flow after tax for ROI criterion.

Year	Capital cost	Operating cost	Interest on Loan at 15%	Sales Income	Cash Flow before tax	Tax Allowance (Depreciation)	Taxable Cash Flow	Corporation Tax at 52%	Net Cash Flow after Tax
0	−996				−996				−996
1	−1295		−149		−1444				−1444
2		−1668	−344	+2749	+737				+737
3		−1668	−344	+2749	+737	189	548	285	+452
4		−1668	−344	+2749	+737	189	548	285	+452
5		−1668	−344	+2749	+737	189	548	285	+452
6		−1668	−344	+2749	+737	189	548	285	+452
7		−1668	−344	+2749	+737	189	548	285	+452
8		−1668	−344	+2749	+737	189	548	285	+452
9		−1668	−344	+2749	+737	189	548	285	+452
10		−1668	−344	+2749	+737	189	548	285	+452
11		−1668	−344	+2749	+737	189	548	285	+452
12	+339*						548	285	+114

* Recovery of scrap value and working capital.

11.4.1 Return on investment (ROI)

This is usually defined as the average annual net cash flow (the surplus remaining after meeting all operating expenses out of income, either before or after tax), divided by the total fixed investment, and expressed as a percentage. Consideration of the net cash flow after tax will be made since the provision for taxation does allow a more realistic assessment to be made. Although this method is extensively used, it possesses inherent disadvantages which are discussed in Section 11.4.2. The capital cost of the plant is assumed to be provided by a loan for which interest is paid at an annual rate of 15% which represents the worst situation as it is likely that part of the investment will be repaid out of receipts during operation of the project. Table 11.4 illustrates calculation of the net cash flow after tax from which the return on investment after tax is given by,

$$\text{ROI after tax} = \frac{737 + (9 \times 452)}{10 \times (1992 - 100)} \times 100\% = 25.4\%$$

When this value is compared to a minimum acceptable rate of return of 20% the project seems attractive but such a small difference merits re-examination of the cash flow data using other profitability criteria.

11.4.2 Net present worth and discounted cash flow rate of return

The ROI after tax simply compares the net income with capital expenditure and takes no account of the time value of money. This can be very important factor in assessing profitability as interest needs to be charged on borrowed capital and surplus income may, at least theoretically, be invested to secure additional income until cessation of the project. Therefore a cash flow which will occur in the future life of a project has a present worth depending upon the interest rate at which it could have been invested to earn between the present time and the time when the cash flow exists. The present worth of a cash flow is determined by multiplication of the latter by the discount factor, $(1+i)^{-n}$ where i is the discount rate and n is the number of years from the present time. The process of calculating all the present worth of incomes and expenditures during the lifetime of a project is known as discounting and the algebraic sum of the present worths of income (positive) and expenditure (negative) is defined as the Net Present Worth (NPW).

For a given discount rate, the NPW should be positive if the project is to be profitable. Since the NPW criterion depends upon the discount rate selected, there will be a value of i for which the NPW is zero and this is known as the Discounted Cash Flow (DCF) rate of return and is the third criterion for profitability used in this study.

A computer program was formulated to evaluate the cash flows and profitability criteria for the base case, that is for the estimated values of capital and operating costs together with the selling price of the MEK product taken as the current market price [17].

Since the determination of the DCF rate of return is a trial and error procedure, iteration using a computer program was preferred since this also permits sensitivity

Table 11.5 — Economic evaluation by NPW and DCF rate of return for the base case × £1000.

Year	Income	Expenses	Cash flow before tax	Tax allowance	Tax payable	Cash flow after tax	Present worth	NPW
0	0.	996.	-996.	0.	0.	-996	-996.	-966.
1	0.	1295.	-1295.	0.	0.	-1295	-1079.	-2075.
2	2749.	1668.	1081.	189.	0.	1081.	751.	-1324.
3	2749.	1668.	1081.	189.	464.	617.	357.	-967.
4	2749.	1668.	1081.	189.	464.	617.	298.	-669.
5	2749.	1668.	1081.	189.	464.	617.	248	-421.
6	2749.	1668.	1081.	189.	464.	617.	207.	-215.
7	2749.	1668.	1081.	189.	464.	617.	172.	-42.
8	2749.	1668.	1081.	189.	464.	617.	144.	101.
9	2749.	1668.	1081.	189.	464.	617.	120.	221.
10	2749.	1668.	1081.	189.	464.	617.	100.	321.
11	2749.	1668.	1081.	189.	464.	916.	123.	444.
12	100.	0.	100.	0.	464.	-364.	-41.	403.

D.C.F. rate of return = 25.8%

analysis using different cash flow data as described in Section 11.5. The output of the program for the base case is given in Table 11.5 and the basic steps may be summarised as follows,

(i) The income is derived from sale of the product assuming a steady demand throughout the project life. The expenditure includes capital investment plus 15% working capital over the first two years together with the operating costs incurred when production commences in year 2, as calculated by equation (11.1). The difference between these represents the cash flow before tax. The working capital is recovered in the final year of production as a cash flow after tax.

(ii) Depreciation of plant is tax allowable and therefore tax is payable on the difference between the cash flow before tax and the annual depreciation. As discussed in Section 11.3.5 taxes are paid one year after they are incurred, in order to give the cash flow after tax.

(iii) The present worth of the cash flow after tax is obtained by discounting. A discount factor was selected to represent a minimum acceptable rate of return and is approximated to the cost of capital taken as 11% plus an inflation rate (9%) with negligible allowance for risk. A discount rate of 20% was used and the associated discount factors were evaluated in the program by the following expression

$$\text{Discount Factor} = \left(\frac{1}{1+i}\right)^n$$

where i is the discount rate = 0.20
n is the year of the project.

The cash flow after tax is multiplied by the discount factor for each year.

(iv) The present worth is cumulated over the project life to determine the NPW of the project.

(v) Finally, the above procedure is iterated until a NPW of zero is obtained by adjusting the value of the discount factor. The final value of the latter represents the DCF rate of return. The logic flow diagram used in the determination of both the NPW and DCF rate of return is shown in Appendix K.

It may be seen from Table 11.5 that the base case situation produced a large positive value of NPW and a DCF rate of return of approximately 26% and from this it may be concluded that the project is an acceptable proposition and likely to be profitable.

11.5 Sensitivity analysis

The majority of the cash flow data used in the above analysis was estimated using correlations based on existing processes and the accuracy of such data is consequently not as high as a detailed estimating procedure that a contractor would employ. Therefore a certain degree of uncertainty is associated with the use of

Table 11.6 – *Project evaluation - sensitivity analysis.*

	1 Base case	2 Capital up 10%	3 Operating up 10%	4 Raw materials up 10%	5 Income down 10 %	6 1 year delay in start up	7 (2+4+5)	8 (2+3+5+6)	9 Capacity double	10 Capacity half
NPW (at 20%) X(£1000)	403	180	73	212	−141	179	−556	−931	1888	−156
DCF Rate of return (%)	25.8	22.2	21.1	23.3	18.6	22.2	12.5	11.3	36.6	16.8

short cut estimating techniques, and sensitivity analysis may be employed to examine the consequences of estimating errors. A relatively arbitrary change is made to each estimated figure to ascertain the effect of this change on the profitability of the project. For this study, the computer program was re-run using the changes outlined in Table 11.6. The results of the analysis show that the profitability is most sensitive to changes in the income from sale of the product. Therefore the demand for the product and its marketing and sales would merit more attention as small changes significantly affect the process profitability. This may be investigated by increasing or decreasing the sales income throughout the project life, if demand fluctuations could be predicted. In practice, however, an established process for the manufacture of a commonly used solvent such as MEK, is unlikely to be uneconomic for technical or marketing reasons and this is proposed as justification for a constant sales income over the project life. As expected, combinations of changes produce more pessimistic situations, but the profitability of their occurence together can be considered much lower. Finally, the economic evaluation was completed for the hypothetical cases where the plant capacity was doubled or halved which illustrates quite clearly the benefits of increasing the scale of an operation provided the product demand and price can be maintained.

11.6 Analysis of alternative process routes

The alternative process routes have been described in Chapter 2 and since they are both technologically feasible, the decision to implement either one is based on economic grounds. Essentially, it is necessary to determine if the installation of a recovery section of plant is justified by the increased yield of product per tonne of feed. The outcome of the latter will be assessed using the incremental ROI criterion. The cost of the system to recover the MEK and 2-butanol from the vapour stream leaving the condenser was calculated using the factor estimating method[48]. This technique expresses the total fixed investment as the product of the delivered equipment cost and a Lang factor to account for direct and indirect costs.

11.6.1 Calculation of equipment cost, I_E

The equipment for the recovery section is shown in Figure 15.1 and their respective costs have been evaluated by various methods and from different sources.

(i) *Absorption Column* – This stainless steel tower is 0.5 m in diameter and 7.0m high, packed with 5.0 m of 3.8 cm Raschig rings (stoneware). Using the cost correlation presented by Peters and Timmerhaus[56], column diameter = 19.7 inches from Figure 15-26[56]. Cost per foot of tower height = \$480 (January 1967). Updating the cost, applying a location index and converting the currency total cost of tower including materials =

$$£ \frac{480 \times 22.96}{1.1 \times 1.8} \times \left(\frac{180}{84.2}\right) = £11\,901$$

(ii) *Solvent Extraction Column* – Clerk[57] has reported costs of rotating disc contactors excluding the cost of the agitator motor and this data will be employed to determine the cost of the stainless steel RDC designed in Chapter 7 to be 0.2 m diameter and 2.5 m height. Diameter of column = 7.87 inches, then Purchase cost[56] = $150 per foot of height (October 1964). Applying a factor of 2.6 for materials of construction (304 stainless steel),

$$\text{Total cost of RDC} = \frac{150 \times 8.202 \times 2.6}{1.1 \times 1.8} \times \left(\frac{180}{79.3}\right) = £3668$$

from Section 7.4.2

Power input to column, $E = 11.93$ W/m^3

Volume of column = $\tfrac{1}{4}\pi(0.02)^2\,2.5 = 0.0785$ m^3

then Power requirement assuming 50% power losses in bearings and gearing
= 11.93 × 0.0785 × 2.0 × 1.34 × 10^{-3} = 2.5 × 10^{-3} hp

The equivalent cost of a 0.25 hp agitator motor is given by Peters and Timmerhaus in Figure 13.54[56] as £216.

∴ Total Cost of Solvent Extraction Column = £3884

(iii) *Solvent Stripping Column* — From Chapter 7, the dimensions of this column for recovery of trichlorethane from the extract phase are 0.44 m diameter, 7.30 m high (plus an allowance of 2.0 m for distributors and headers) packed with 2.5 cm Raschig rings. From Figure 15-26[56] cost per foot of height = $320 (January 1967).

$$\text{Total cost} = £\frac{320 \times 30.5}{1.1 \times 1.8} \times \left(\frac{180}{84.2}\right)$$

Cost of Stripping Column = £10 539.

(iv) *Hydrogen Drying Equipment* — If the MEK and 2-butanol are recovered from the vapour leaving the condenser, the hydrogen will be of sufficient purity to be employed in the regeneration of the brass catalyst in the tubular reactors, provided that it is dried to remove water vapour taken up in the absorber. Drying will be accomplished by passing the gases from the absorber through a dehumidification column through which water at 4°C is pumped countercurrently over a packing. Preliminary estimates indicate that a 0.34 m diameter column, 8.0 m high, containing 2.5 cm Raschig rings, will remove most of the water. Using the same cost correlation as for the absorption column, from Figure 15-26[56]

Cost per foot of height = $140 (January 1967)

$$\text{Cost of dehumidification column} = £\frac{140 \times 26.25}{1.1 \times 1.8} \times \left(\frac{180}{84.2}\right)$$

$$= £3968$$

Since cooling water is available at 24°C, it is necessary to install a refrigeration unit for which an approximate cost was determined using the data given by Rudd and Watson[15]

$$\text{Cost of refrigeration unit} = £\frac{4200}{1.1 \times 1.8}\left(\frac{180}{77.6}\right)$$

$$= £4919$$

(v) *Heat Exchangers* – There are four heat exchangers in the recovery plant and their respective costs were determined by the correlation presented by Bridgwater[48] and also using the ratio method presented by Russ and Watson[15].

For the absorber liquor cooler, heat transfer area = 16.23 m² (174.7 ft²)

for carbon steel cost = $533 (16.23)^{0.57}$ (January 1977)

Updating and multiplying by a factor of 2 for stainless steel material of construction.

$$\text{cost} = £533(16.23)^{0.57} \times 2 \times \left(\frac{300}{285}\right)$$

$$= £5495$$

Using the ratio method [15]

$$\text{cost} = 1350 \left(\frac{174.7}{50}\right)^{0.48} \text{ (1961) USA}$$

$$\therefore \text{cost} = £\frac{1350 \times 2}{1.1 \times 1.8} \left(\frac{174.7}{50}\right)^{0.48} \left(\frac{180}{77.6}\right)$$

$$= £5764$$

Taking the mean of these values,

    Cost of absorber liquor cooler = £5630

Similarly for the other heat transfer equipment,

    Cost of solvent cooler = £2007
    Cost of stripping column reboiler = £1912
    Cost of stripping column condenser = £1597
    Total cost of Heat Exchangers = £11146

(vi) *Storage Tanks* – Considering the extract storage tank; assuming that 24 hour storage capacity is required,

    Total mass to be stored = 1126 × 24 = 27024 kg
    Density of extract = 1220 kg/m³
    Tank volume required, V = 22.15 m³

Using the equation presented by Bridgwater[48] for stainless steel tanks,

$$C = 222(V)^{0.61} \text{ (January 1977)}$$

$$\text{Cost of extract storage tank} = 2 \times 222(22.15)^{0.61} \left(\frac{300}{285}\right)$$

$$= £3093$$

similarly

  Cost of Raffinate storage tank = £4830
  Cost of Solvent storage tank = £ 119
  Total cost of storage tanks = £8042

(vii) *Pumps* — The cost of the four transfer pumps was determined using the following equation[48] for a general purpose centrifugal pump fabricated from stainless steel

  Cost of pump including motor = $29.3 (G)^{0.4}$ (January 1977)
  where G is the capacity in gallons/hour

  Cost of absorber liquor pump = £390
  Cost of raffinate pump = £378
  Cost of extract pump = £278
  Cost of solvent pump = £246
  Total cost of pumps = £1292

(viii) *Blower* — It is proposed to install a blower in order to transfer the hydrogen through the drying column to the reactors.

  Hydrogen flowrate = 37.5 kg/h
       = 247.5 ft^3/min

For a rotary blower of this capacity, from Figure 13-50[56]

Cost = $ 1600 (January 1967)

Allowing a factor of four for an explosion-proof blower and updating the cost,

$$\text{Cost of blower} = £\frac{1600 \times 4}{1.1 \times 1.8} \times \left(\frac{180}{84.2}\right)$$

$$= £6910$$

The total delivered equipment cost is summed in Table 11.7.

Table 11.7 — *Delivered equipment cost for MEK recovery plant.*

Plant Equipment	Cost £
Absorption column (+ internals)	11901
Solvent extraction column (+ internals)	3884
Solvent stripping column (+ internals)	10539
Hydrogen drying equipment	8887
Heat exchangers	11146
Storage tanks	8042
Pumps and Hydrogen Blower	8202
Total	62601

11.6.2 *Total fixed investment*

This is calculated by the use of Lang factors which express the remaining direct and indirect costs as a function of the equipment cost[48], E. The values used for this fluid processing plant are given in Table 11.8 and the Lang factor was found to be 4.18.

Then Total Fixed Investment = £4.18 × 62601
= £261 672

Table 11.8 — *Derivation of Lang factor for MEK recovery plant.*

	Range	Typical value for Fluid processing plant	Percentage of E
Direct costs			
Deliver Equipment, E	E	100	100
Equipment Installation	$0.18-1.0E$	47	47
Instrumentation and Controls (installed)	$0.06-0.20E$	18	18
Piping (installed)	$0.09-0.6E$	66	66
Electricals (installed)	$0.06-0.35E$	11	11
Buildings	$0.15-1.2E$	18	18
Site Improvements	$0.04-0.25E$	10	10
Service Facilities (installed)	$0.1-1.0E$	70	20
Land (on existing site)	$0.02-0.1E$	6	0
Total Direct Plant Cost	$0.7-4.7E$	346	290
Indirect costs			
Engineering and Supervision	$0.15-0.6E$	33	33
Construction Expenses	$0.15-0.6E$	41	41
Total Direct and Indirect Costs, X	$1.0-5.9E$	420	364
Contractors' Fee	$0.05-0.15X$	21	18
Contingency	$0.01-0.15X$	42	36
Fixed Capital Investment		483	418
Lang Factor		4.83	4.18

11.6.3 *Profitability assessment*

The advantages of installation and operation of the MEK recovery plant are increased income from additional MEK and 2-butanol plus an effective income from the value of the hydrogen. This revenue arises because if the organic components in the vapour leaving the condenser were not removed then the gases would be burned to raise steam. However, hydrogen leaving the absorber is dried to a condition where it is suitable for use in the catalyst regeneration process and therefore the recovery plant produces useful hydrogen which would otherwise be produced on a separate plant. If the gases were burned to produce steam at 3.6 bar.

Heat availability = 15.628×10^{10} kJ/year

Assuming 80% boiler efficiency using a feed of water at 293 K,

Steam produced = 47200 tonnes/year

Value of this steam if the cost of steam raising is £8/tonne is given by £47200 × 8

Cost of Steam = £377 603

An approximate value of the capital cost for a steam reforming plant producing hydrogen for catalyst regeneration at a rate of 247.5 ft^3/min is[58] £0.45×10^6

For the recovery plant,

Value of MEK produced at the current selling price

$$= 212 \times 0.2545 \times 8000$$
$$= £431\,632$$

Value of 2-butanol produced at the feedstock cost

$$= 14 \times 0.0777 \times 8000$$
$$= £8705$$

∴ Total value of recovered organics = £440 337

The result of installation of the hydrogen producing plant, when the vapours leaving the condenser are burned, is now economically assessed using the incremental return on investment criterion which is defined as,

$$\text{inc ROI} = \frac{\text{average additional annual income}}{\text{additional fixed investment}} \times 100\%$$

$$= \frac{(\text{Steam Raising Cost}) - (\text{Cost of recovered organics})}{(\text{Cost of hydrogen plant}) - (\text{Cost of Recovery Plant})} \times 100\%$$

$$= \frac{(377\,603 - 440\,337)}{(450\,000 - 261\,672)} \times 100\%$$

$$= -33.3\%$$

Since a negative result is obtained, then it may be concluded that burning of the vapours and installation of a hydrogen producing plant is uneconomic and therefore recovery of the MEK and 2-butanol is shown to be the better alternative.

Chapter 12

OPERABILITY STUDY OF THE REACTOR SECTION

12.1 General discussion

An "operability study" is a structured technique for identifying potential mal-function beforehand. Critical examination in process design, inevitably leads to a safer, more reliable and more profitable process[59]. These studies are based on sets of "key words" which stimulate thought. It is essential to formulate unambiguously, the design objectives, to which the key words are applied in a systematic way.

In operability studies the thoughts are about deviations from the design conditions and the design intention. The set of key words contains two sub-sets: "property words" which focus attention on the design conditions and the design intention; and the second sub-set of "guide words" which focus attention onto possible deviations. The Chemical Industries Association[60], has published a *Guide to Hazard and Operability Studies* which recommends the seven guide words explained

Table 12.1 – *A list of guide words*[2].

Guide Words	Meanings	Comments
NO or NOT	The complete negation of these intentions	No part of the intentions is achieved but nothing else happens
MORE, LESS	Quantitative increases or decreases	These refer to quantities and properties such as flow rates and temperatures as well as activities like "HEAT" and "REACT".
AS WELL AS	A qualitative increase	All the design and operating intentions are achieved together with some additional activity
PART OF	A qualitative decrease	Only some of the intentions are achieved; some are not
REVERSE	The logical opposite of the intention	This is mostly applicable to activities, for example reverse flow or chemical reaction. It can also be applied to substances, *e.g.* "POISON" instead of "ANTIDOTE" or "D" instead of "L" optical isomers
OTHER THAN	Complete substitution	No part of the original intention is achieved. Something quite different happens.

in Table 12.1. This set is used here, but the set is not mandatory; it may be more efficient to use only NO, MORE and LESS, with a layer set of property words. However, it is essential not to overlook meaningful deviations and the key words should be chosen beforehand accordingly.

Recognising the potential causes of deviations, when using only a process and instrumentation diagram, requires some industrial experience. Engineers who may lack this experience will find it useful to read Chapter 4 of *Flowsheeting for Safety*[61], and to peruse issues of the *Loss Prevention Bulletin* published by the Institution of Chemical Engineers, for which cross-referenced indexes are available.

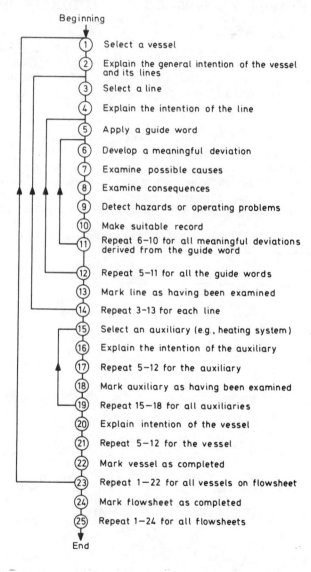

Figure 12.1 — *Detailed sequence of an operability study.*

The 1978 Jubilee lecture by Kletz to the Society of Chemical Industry[62] contains many useful guide-lines for designing intrinsically safe processes, which should be referred to at an early stage in the design project.

12.2 Operability studies

These are generally carried out in industry by a small team of specialists, directed by an experienced group-leader. The exercise is of the brain-storming type and only those deviations which would lead to hazardous outcomes are recorded for further action. Initially the inexperienced chemical engineer will find it more effective to record all causes of deviations, their consequences and action. In addition it is more appropriate to use a process flowsheet with only the basic control loops. This enables those who are not experienced instrument technologists, to identify the need for additional instruments which are included on the final diagram. A further benefit of keeping complete records of the operability study, is that operating instructions and procedures for start-up, shut-down and carrying out cyclic operations can be drawn up by reference to the operability study. Furthermore, if the significance of deviations in measured variables is understood, simple fault-finding strategies can be included in the design report. The operability study in this design project differs from that of an industrial process design in that normally every vessel and pipeline must be covered by an industrial operability study. A typical sequence for carrying out these procedures is shown on Figure 12.1. In this design report the operability study is restricted to the reactor section described in Chapter 4. A "truncated operability study" of this type may introduce difficulties, because deviations originating upstream of the "truncation line" can only be specified in general terms. Nevertheless, the principles of the analysis can be introduced, illustrating that it is possible to perform a truncated operability study effectively. The reactors are designated 1, 2, 3, on the operability study flowsheet, Figure 12.3. Line numbers are bracketed.

12.3 Preparation for the operability study

Figure 12.2 shows a typical 15 hour period of the utilisation schedules for the three reactors. Four important operating periods are identified, during which different operations are being carried out in two of the three reactors while the third is on stand-by. These periods are designated A to D as follows; where time t is expressed in hours from the datum on Figure 12.2

$A : 1 < t < 6$
$B : 6 < t < 6.5$
$C : 6.5 < t < 11.5$
$D : 14.5 < t < 15$

The following property words were drawn up by reference to the description of the design conditions and design intentions of the three reactors: FLOW. TEMPERATURE. PRESSURE. DEHYDROGENATE. OXIDISE. PURGE. REDUCE. CYCLE-TIME. PREHEAT. STAND-BY. CONCENTRATION.

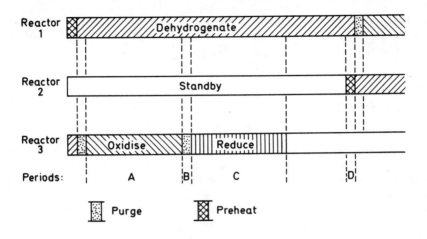

Figure 12.2 — *Operating periods in a cycle.*

12.3.1 *Mechanical design details of the reactors*

Each reactor consists of a vertical tubular heat exchanger. The tubes can withstand 900 K with a pressure difference of 2.5 bar. The tubes are identically filled with brass pellets and the central tube in the bundle has a number of temperature indicators attached to points along its length.

12.3.2 *Design intention and design conditionss,*

12.3.2.1 Period A

Reactor 1 To dehydrogenate 2-butanol vapour to MEK over brass catalyst pellets. The vapour enters the lower header at 773 K and 2.5 bar. The total flow of 1558.5 kg/h is equally distributed between the tubes. Products with 10% unreacted feed, leave the top of the reactor at 642 K and 1.7 bar. Flue gases at 4124 kg/h enter the shell at the top of the exchanger at 800 K and leave the bottom of the shell at 780 K. The concentration of O_2 in the gases is 1.3% w/w and there is no carbon monoxide.

Reactor 2 On stand-by. No feedstock and no flue gas are present. The reactor tubes contain nitrogen only, and fully regenerated catalyst.

Reactor 3 To oxidise fouling deposits on the surface of the brass catalyst pellets with air entering at 700 K and atmospheric pressure. The oxidation is exothermic. Oxidation cycle time is 5 hours.

12.3.2.2 Period B

Reactors 1 and 2 Conditions all the same as during Period A.

Reactor 3 Oxygen is purged from the tubes by admitting nitrogen at 700 K and atmospheric pressure for a cycle time of half an hour.

12.3.2.3 Period C

Reactors 1 and 2 Conditions are the same as during Period A.

Reactor 3 The catalyst is reduced by the exothermic reaction of its oxidised surface, with hydrogen entering at 700 K and atmospheric pressure. The cycle time is 5 hours.

12.3.2.4 Period D

Reactor 1 Conditions are the same as during Period A.

Reactor 2 Preheated to 700 K by flue gases entering the shell at 800 K. The cycle time is half an hour.

Reactor 3 On stand-by. Conditions are the same as for Reactor 2 during Period A.

12.4 Discussion of operability study

The operability study which is presented in Table 12.2 has revealed the need for some additions and modifications to the equipment shown on the operability study flowsheet, Figure 12.3. These modifications arise principally because dehydrogenation is carried out at high pressure, whereas the regeneration operations occur at atmospheric pressure. This pressure differential could lead to reverse flow in the service lines 15, 16 and 19, should some valves be leaking or left

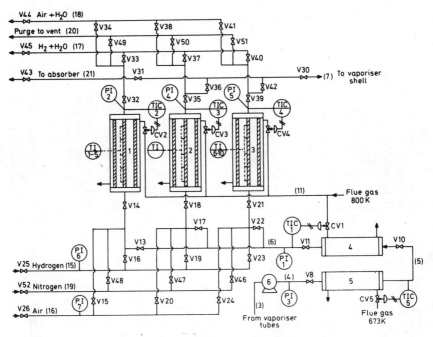

Figure 12.3 — *Operability study flowsheet.*

Table 12.2 – Operability study for period A (Refer to operability study flow sheet for line identification).

Line 6 To convey 2-butanol vapour at 773K and 2.5 bar from the second stage vapour superheater 4 to reactor 1 at 1558.5 kg/h.

Property Word	Guide Word	Cause	Consequence	Action (*indicates minor modification)
Flow	No	1) Catalyst bed in 1 is blocked.	1-6) Temperature indicators TI 1-5 are all equal to the flue gas temperature (800K) and TIC2 shows low.	1) Transfer to Period D operation.
		2) Compressor 6 is failed.		2) Put plant on standby while compressor 6 is repaired.
		3) V13 OR V14 closed.	1-6) Vaporisation in thermosyphon boiler stops.	3-5) Check for incorrect closure or blockage of valves, noting PI1, PI2 and PI3 to diagnose fault.
		4) V32 OR V31 closed.	1-6) No flow in line 7.	
		5) V11 closed.	3) PI1 = PI3 AND PI2 shows low.	6) Check line 5, flow, no.
		6) No flow in line 5.	4) PI1 = PI2=PI3.	
			5) PI1=PI2 both low AND PI3 high.	
	More	1) More flow in line 5.	1) More flow in line 11.	*2-4) Fit non-return valves on the downstream side of each of the following isolation valves; in order to avoid contamination of services: V15, V16, V19, V20, V23, V24, V46, V47, V48.
		2) V16 AND 425 open or leaking.	2) Hydrogen supply is contaminated with butanol at 2.5 bar and 773K.	
		3) V15 open or leaking	3) Air supply is stopped, reverse flow in line 16, butanol flows through line 7 to line 18.	
		4) V48 AND V52 open or leaking.	4) Nitrogen supply is contaminated with butanol at 2.5 bar and 773K.	
	Less	1) Tubes in 4 are leaking.	1) Butanol vapour enters flue gas.	1) Stop 6, close V10 and V11, plug tubes in 4. *Could 4 be bypassed?
	As well as	1) Tubes in vaporiser are leaking.	1) MEK and hydrogen enter lines 3, 4, 5 and 6.	*1) Fit hydrogen detector in line 3 with an alarm.
Temperature	More	1) TIC1 controlling high.	1-2) Temperature indicator TI5 at bottom of 1 shows high.	1) Reduce set point of TIC1.
		2) CV1 stuck open.		2) Overhaul CV1.
	Less	1) Superheater 4 is fouled.	1) CV1 wide open AND TIC1 shows low.	1) Increase set point of TIC5 or reduce feed rate.
		2) TIC1 controlling low.	1-3) Temperature indicator TI5 at bottom of 1 shows low.	2) Increase set point of TIC1
		3) CV1 stuck closed.	1-3) Less dehydrogenation in 1.	3) Overhaul CV1.
Pressure	More	1) V13 OR V14 partially closed.	1,2) PI2 shows low AND PI shows high.	2) Transfer to period D operation.
		2) Catalyst bed in 1 is partially blocked.	3) PI2 shows high.	1-3) Check pressures to locate cause of more pressure.
		3) V32 OR V31 partially closed.		
	Less	1) V11 partially closed.	1,2) PI1 AND PI2 shows low and PI3 shows high.	1) Check V11 is fully open
		2) Tubes in 4 are partially blocked.		*2) Could 4 by partially bypassed?

Line 16 *To introduce air at 700 K and atmospheric pressure to reactor no. 3.*

Property Word	Guide Word	Cause		Consequence		Action (*indicates minor modification)
Flow	No	1) Catalyst bed in 3 is blocked. 2) V26 OR V24 OR V21 closed. 3) V39 OR V41 OR V44 closed.	1-3) 1-3)	Temperature indicators TI6-10 in 3 are all equal to TIC4, all show low. Oxidise operation cannot proceed; cycle time will be more than 5 hours.	1) 2,3)	Monitor pressure drop across the dehydrogenating reactor. Switch to standby reactor if the pressure drop exceeds 1 bar. Check that valves 21, 24, 26, 39, 41, and 44 are open.
	Less	1) Catalyst bed in 3 is heavily fouled. 2) Partial blockage of V26 OR V24 OR V21 OR V39 OR V41 OR V44.	1-2)	Temperature indicators above hot spot in 3 show low. Hot spot moves slowly. Cycle time exceeds 5 hours.	1,2) 2)	Raise set point of TIC4 to allow CV4 to admit more flue gas to the shell. Monitor temperature indicators TI6-10 in 3 to ensure that hot spot temperature does not exceed 1000K. Check that valves 21, 24, 26, 39, 41 and 44 are fully open.
	As well as	1) V15 leaking. 2) V22 leaking. 3) V25 AND V23 open or leaking.	1,2) 3)	Butanol vapour enters line 7 and escapes through line 18. Hydrogen enters 3. EXPLOSION.	*1) 2) *3)	Fit non-return valves. See line 6, flow, more. Place plant on standby while V22 is replaced. Fit pressure gauges PI6 and PI7. Ensure that PI6 < = PI7 AND V24 is fully open.
Temperature	More	1) Air preheater too effective.	1)	Hot spot temperature in 3 exceeds 1000K.	1)	Reduce air flow by reducing V21.
	Less	1) Air preheater is fouled OR heating medium has low flow OR low temperature.	1)	Temperature indicators below hot spot in 3 show low.	1)	Raise set point of TIC4 to allow CV4 to admit more flue gas to the shell. Hot spot must not exceed 1000K.
Pressure	More	1) Catalyst bed in 3 is heavily fouled. 2) Partial blockage of V24 OR V21 OR V39 OR V41 OR V44.	1,2)	Less flow in line 16 AND line 18 AND PI7 shows high.	1,2) 2)	Check line 16, flow, less. Isolate 3 and overhaul the faulty valves.
	Less	1) V26 partially blocked.	1)	Less flow in line 16 AND PI7 shows low.	1)	Close V24 and overhaul V26. Max time available for overhaul or replacement of V26 is 5 hours.

173

Property Word	Guide Word	Cause	Consequence	Action (*indicates minor modification)

Reactor 1. *During period A its function is to dehydrogenate 2-butanol vapour to MEK. Design conditions have been detailed in Section 12.3.3.*

Property Word	Guide Word	Cause	Consequence	Action (*indicates minor modification)
Flow	More	1) More flow in line 5.	1) Concentration of butanol in line 7 exceeds 10.9% w/w.	1) Reduce flow through V11.
	Part of	1) Some catalyst packed tubes become blocked.	1) PI1 shows high AND concentration of butanol in line 7 exceeds 10.9%. 1) Blocked tubes overheat and bend.	1) Transfer to period D operation. *1)Inspect tubes if fouling has been severe.
	Less	1) Less flow in line 6.	1) More than 90% dehydrogenation of butanol.	1) Check line 6, flow, less.
Dehydrogenate	More	1) Less flow in line 6. 2) TIC2 controlling high.	1) Temperature indicators TI1-5 in 1 show high AND flow in line 6 is normal.	2) Reduce set point of TIC2.
	Less	1) Catalyst pellets are fouled.	1) Concentration of butanol in line 7 exceeds 10.9% w/w	1) Increase set point of TIC2 AND reduce flow through V11 OR transfer to period D.
Temperature	More	1) TIC2 controlling high.	1) Temperature indicators TI1-5 in 11 show high AND flow in line 6 is normal.	Reduce set point of TIC2.
	Less	1) Temperature of flue gas is less than 800K. 2) Tubes side in 1 are fouled. 3) Shell side in 1 is fouled.	1) TIC2 shows less than 642K AND CV2 is wide open AND temperature indicators TI1-5 show low. 2) TIC2 shows less than 642 K AND CV2 is wide open AND temperature indicators TI1-5 show high.	1) Increase flue gas temperature. 2) Transfer to period D and prolong oxidation. 3) Planned maintenance is required to clean the shell side of the reactors.
Cycle time	Less	1) Premature fouling of catalyst caused by TIC2 controlling high OR mal-distribution of feed to the reactor tubes.	1) Dehydrogenation diminishes.	1) Transfer to period D.

Property Word	Guide Word	Cause	Consequence	Action (* indicates minor modification)

Reactor 2 *During period A this reactor should be on standby.*

Standby	Other than	1) V25 AND V19 AND V35 AND V37 AND V45 have been left open after reduction of catalyst. 2) CV3 leaking.	1) Hydrogen flows through 2 and line 17 contains pure H$_2$. 2) Flue gas flows through the shell of reactor 2.	1) Check that valves 18,19,25,35,37 and 45 are shut. *2) Fit block valves as well as control valves on the flue gas supply lines to each reactor. Close block valve on reactor 2.

Reactor 3 *During period A the fouling deposits in this reactor are being oxidised. Design conditions have been detailed in Section 12.3.3.*

Temperature	More	1) TIC4 controlling high. 2) CV4 stuck open. 3) Flow in line 16 high.	1-3) Hot spot temperature shown by TI6-10 exceeds 1000K. Tubes bend due to thermal stress. 3) Hot spot becomes an extended hot zone along the tubes.	1) Monitor TIC4 and TI6-10 continually, Hot spot temperature must not exceed 1000K. Reduce set point of TIC4. 2) Reduce flue gas flow by means of *manual valve on flue gas line to reactor 3. Overhaul CV4 when reactor 3 goes onto standby after period C. 3) Reduce flow through V21.
	Less	1) Oxidation of deposits is nearly complete. 2) TIC4 controlling low. 3) CV4 failed closed.	1) TI6-10 show no hot spot. Water vapour in line 18 falls. 2,3) Oxidation requires more than 5 hours. TI6-10 show a hot spot. which moves slowly.	*1) Install a hygrometer to be shared between lines 17 and 18, in order to monitor the progress of oxidation and reduction. 2) Raise set point of TIC4. 3) Close *manual valve on flue gas line to reactor 3 and overhaul CV4. Max. time available for overhaul or replacement of CV4 is 15 hours.
Pressure	More	1) V22 is open or leaking	1) PI5 indicates more than 1 bar. Butanol vapour lost through line 18.	1) Close V11 and put plant on standby while V22 is replaced. Purge with nitrogen before continuing oxidation.
Oxidise	Less	1) Low flow in line 16.	1) Cycle time exceeds 5 hours.	1) See line 16, flow, less.
	More	1) High temperature in reactor 3 shell	1) High hot spot temperature. Tubes become overheated and bend.	1) See reactor 3, temp., more.

Property Word	Guide Word	Cause	Consequence	Action (*indicates minor modification)
	Other than	1) V22 is open or leaking	1) PI5 shows high. If butanol vapour dehydrogenates, the resulting hydrogen/air mixture may explode.	1) Close V11, stop heating reactor 3, replace V22 and purge with nitrogen before continuing oxidation.

Line 7 *To transfer products at 1558.5 kg/h to the shell of the thermosyphon vaporiser. Products are at 1.7 bar and 642 K and have the following composition by weight: MEK 86.7%, hydrogen 2.4%, butanol 10.9%.*

Property Word	Guide Word	Cause	Consequence	Action (*indicates minor modification)
Flow	No	1) No flow in line 6. 2) V30 is closed.		1) See line 6, flow, no.
	Less	1) Less flow in line 6. 2) Tubes in reactor 1 are leaking. 3) V42 is open or leaking. 4) V36 AND V38 are open. 5) V36 AND V50 OR V37 are open. 6) V43 is open or leaking.	1-6) Vaporisation in boiler diminishes. 2) Products are lost with flue gases. 3-4) Products are lost via line 18. 5) Products are lost via lines 20 OR 17 respectively. 6) Products flow to the absorber.	*2-6) Monitor lines 17, 18, 20 and 21 for MEK (smell?)
Temperature	Less	1) TIC2 controlling low OR CV2 stuck. 2) Low flow and high heat loss.	1,2) Vaporisation in boiler diminishes.	1) See reactor 1, temp. less.
Concentration of butanol	More	1) Less dehydrogenation.	1) Production of MEK is less.	1) See reactor 1, dehydrogenate, less

Line 18 *To discharge air and water vapour from reactor 3 to a vent.*

Property Word	Guide Word	Cause	Consequence	Action (*indicates minor modification)
Flow	Other than	1) V42 is open or leaking 2) V22 is open or leaking	1) No air enters reactor 3, but products are lost via line 18. 2) No air enters reactor 3, but butanol is lost via line 18.	*1) Monitor line 18 for MEK. *2) Monitor line 18 for butanol.
Oxygen	No	1) No flow in line 16. 2) Low flow in line 16.	1) TI6-10 are all equal to TIC4 which is equal flue gas temp. 2) Diminished hot spot.	1) See line 16, flow, no. 2) See line 16, flow, less.
Water Vapour	No	1) Oxidisation is complete. 2) No flow in line 16.	1) TI 6-10 are all equal to TIC4 which is equal flue gas temp.	*1) Install hygrometer in line 18. 2) See line 16, flow, no.

open. Therefore non-return valves are recommended as shown in Table 12.3. There could be in addition, the possibility of an explosion if hydrogen leaked into the air stream going to Reactor 3. Thus the pressure indicators PI6 and PI7 shown on Figure 12.3 are recommended; with the provision that PI6 must not exceed PI7. This provision would be reversed during Period C. It must not be assumed that valves will never leak nor that they will be completely closed after a given operation. Neither should it be assumed that heat exchanger tubes will not leak. Therefore, detectors have been recommended to aid the operators to recognise if a line contain material which should not be there. Similarly, a hydrometer has been recommended for sharing between lines 17 and 18, to enable the progress of reduction and oxidation to be monitored since both operations produce water vapour. A number of operator actions have also been recommended in the last column of Table 12.2; these have not been abstracted into fault diagnosis strategies and corrective sections; but they could be. However, with so many valves on the flowsheet it is appropriate to list their required states for operating period A. This is done in Table 12.4.

Table 12.3 — *Additional equipment required.*

Equipment	Location	Purpose
9 Non-return valves	Downstream of the following block valves: 15, 16, 19, 20, 23, 24, 46, 47, 48.	To prevent butanol flowing into the service lines in the event of an isolation valve being left open or leaking
1 hydrogen detector	Line 3 before compressor 6	To sound an alarm if tubes in vaporiser leak
2 pressure gauges	PI6 and PI7 on Fig. 12.3	To check that service valves V25 or V26 have not been left open or are leaking. An explosion would result.
3 valves	Flue gas branch lines to the three reactors.	To facilitate manual isolation of reactors not requiring heat
MEK analyser and alarm	Connected to lines 17, 18, 20, 21	To detect loss of product through valves which should be shut
Butanol analyser and alarm	Connected to lines 17, 18, 20.	To detect loss of feed-stock through reactor which is being regenerated.

In performing the operability study it was assumed that the multipoint temperature indicators TI 1-5 in reactor 1 and TI 6-10 in reactor 3 would indicate the typical temperature profile along all other tubes in the respective reactors. This assumption will be invalidated if either the flowrate through the centre tube or the condition of catalyst in this tube, differs from others in the bundle. Although the tubes are of heat resistant alloy, differential expansion between tubes of different wall temperatures, could cause considerable bending. It would be advisable to inspect the shell side of the reactors especially following a period excessive internal fouling.

Table 12.4 — *Required states of isolation valves during period A.*

Open Valves	In Line	Closed Valves	In Line
8	4	17, 18, 22	6
10	5	16, 19, 23, 25	15
11, 13, 14	6	46, 47, 48, 52	19
26, 24, 21	16	15, 20	16
30, 31, 32	7	35, 36, 42	7
39, 41, 44	18	33, 37, 40, 45	17
		49, 50, 51	20
		34, 38	18
		Additional valve in branch to reactor 2.	11

Chapter 13

INSTRUMENTATION AND CONTROL

13.1 General discussion

It is proposed that most of the plant equipment in the MEK process be operated using automatic control with the indicating instruments being located in a central control room. This is the general practice for a plant of this type which is not labour intensive. With the exception of the reactor system, the plant operates at atmospheric pressure and therefore the instrumentation and control will be based upon temperature, flow and level measurements.

Measurements of these parameters will be made using thermo-couples (Chromel-Alumel type), orifice plates and float type indicators respectively. Regulation by pneumatic control is recommended due to the high flammabilities of the process fluids. The final instrumentation scheme is presented in Figure 15.1, but consideration to the principal unit operations will now be given.

13.2 Thermo-syphon vaporiser

Although the reaction products will be employed to vaporise the 2-butanol feed, start-up of the process requires condensing steam as a heating medium on the shell side of the vaporiser. The vaporisation rate is controlled by the steam pressure in a conventional way as shown in Figure 13.1. Immediate transfer to utilise the hot gas stream will result in contamination of the reaction products by vaporised condensate. Since water forms azeotropic mixtures with both MEK (11% w/w water) and 2-butanol (32% w/w water) [63], which would create malfunctioning of the main product distillation column, the hot gases leaving the thermo-syphon

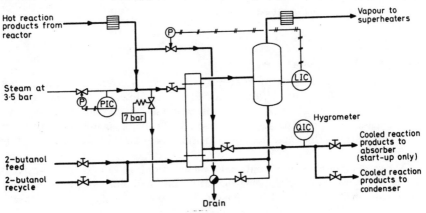

Figure 13.1 — *Control of thermo-syphon vaporiser.*

179

vaporiser will be discharged directly to the absorber. When a hygrometer (electric resistance type), installed on the hot gas exit line, indicates a minimal water content the vapour flow will be diverted to the condenser.

Control of the liquid recirculation rate is determined by a pressure balance over the vaporiser system as discussed in Section 4.10.3 and provided that the liquid level in the knockout drum is maintained, further instrumentation is unnecessary. The liquid level will be controlled by the vaporisation rate in the tubes and since the heat transfer coefficient on the shell side is controlling, the level will be adjusted by control of flow of the reaction products in a by-pass line to the condenser.

13.3 Tubular catalytic reactors

The three catalytic reactors operate in a cycle of 43.5 hours involving "on-line" reaction of 2-butanol, hydrogen reduction and air oxidation of the brass catalyst with intermediate purging using inert nitrogen and "stand-by" periods. If an automatic control scheme were devised for reactor operation with no manual intervention, it would be extremely expensive due to the large number of valves to be actuated; also if any malfunction occurred, then, as discussed in the operability study on the reactors (Chapter 12), the risk of explosion is high. Consequently, manual operation of the reactor operating cycle is proposed with automatic control devices being employed for monitoring of the "steady-state" periods of reaction and catalyst regeneration. The basic instrumentation for the reactors is illustrated in Figure 12.3 and this has been modified as a result of the truncated operability study by the addition of analysis and control equipment described in Table 12.3.

The dehydrogenation reaction is endothermic and the reactors should exhibit a marked degree of self-regulation regarding temperature control[64]. Control of feed temperature is necessary to ensure an adequate reaction rate and this is easily facilitated by manipulation of the flue gas flowrates and temperatures (by blending flue gas from different sources — see Section 4.11) in the two vapour superheaters. The longitudinal temperature profile and fractional conversion in the catalyst-filled tubes are very sensitive to the overall heat transfer coefficient as discussed in Section 4.8. Thus, since the flowrate of reactants inside the tubes is fixed, control may be achieved by variation of the flue gas flowrate to the shell side. Measurement of fluid temperature along the length of the reactor will be made by thermo-couples inserted inside one of the tubes.

13.4 Control of condenser

Under conditions of constant condensate temperature, the heat transfer rate is entirely dependent upon the flow of cooling water. However, a certain amount of subcooling always takes place which varies with the flow of vapour, so that control of condensate temperature is not recommended. A thermo-couple should be placed on the condensate line and the liquid analysed periodically to ensure satisfactory performance.

The most effective way of controlling a condenser is to vary its heat transfer area which is accomplished by allowing the MEK and 2-butanol condensate partially to flood the shell side, thereby reducing the surface available for condensation. Measurement of vapour pressure is a sensitive, indirect means of controlling liquid level, but since the vapour feed to the condenser contains the non-condensible gas, hydrogen, variations in pressure could be due to fluctuations in vapour feed rate and this method is rejected [64]. Therefore, for this partial condenser, direct manipulation of the liquid level, although much slower than vapour pressure measurement, will be adopted and the flow of condensate will be regulated by a level controller.

The cooling water rate will be controlled depending upon the output from an orifice plate placed on the vapour line to the absorption column as shown in Figure 13.2.

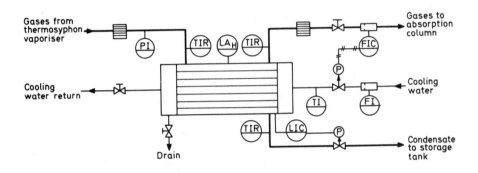

Figure 13.2 — *Control of partial condenser.*

13.5 Absorption column

Since the vapour flow to the absorber is fixed, control of the absorption process will be accomplished by maintenance of the specified liquid/vapour flow ratio. Control of this parameter is preferred because this ratio determines the exit compositions and, as discussed in Chapter 3, the MEK concentration in the liquid should not exceed 11% w/w to avoid formation of an isopicnic in the extraction column. Also this column operates at just over the loading point under steady state conditions, thus providing sufficient flexibility for control purposes. However, during the start-up period, the maximum capacity of the absorption column is required when all the reaction products are transferred directly to the column, firstly from the reactors until the required conversion is achieved, and, finally from the thermo-syphon vaporiser until all traces of steam condensate have been removed. Measurement of the vapour flowrate permits control of this ratio by adjustment of the raffinate flowrate which, in turn, is achieved by regulation of water make-up to the raffinate storage tank using a level controller.

The temperature of the entering liquid also affects the efficiency of the absorption process, but since the exit liquor from the absorber and the recovered solvent from the stripping column are cooled before contacting in the extraction column, the latter operates isothermally at ambient temperature and therefore no temperature regulation is required. Thermo-couples will be inserted on both vapour and liquid inlet streams.

13.6 Extraction column

The solvent/feed ratio in the RDC column will be controlled in a similar way as the liquid/vapour ratio in the absorption column. In this case, flow regulation of 1,1,2 Trichlorethane will be used to compensate for any feed rate fluctuations.

The angular velocity of the rotors will be measured by a tachometer and the set point of the controller will be determined by the drop sizes required in the column.

The extract flowrate will be determined by control of the position of the liquid/liquid interface where the solvent dispersion band coalesces into the bulk extract phase at the base of the column. The location of the interface could be detected by the bouyant force acting on a displacement device.

Efficiency of the extraction operation and the feed composition to the stripping column will be monitored using a refractometer (refractive index MEK at $20°C = 1.3788$; refractive index 1.1.2 Trichlorethane at $20°C = 1.4711$)[63] installed on the extract stream. The control scheme proposed for the absorber and extractor is illustrated in Figure 13.3.

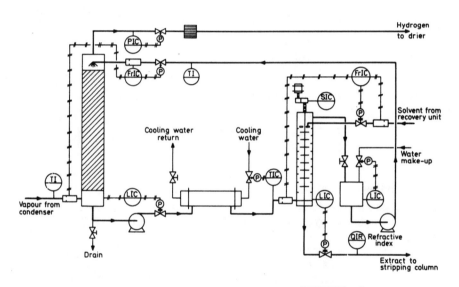

Figure 13.3 — *Control of absorption and extraction units.*

13.7 Solvent stripping column

For continuous, versatile control of distillation processes, the requirement generally adopted is one of constant temperature profile in the column[65]. This ensures preservation of equilibrium between the liquid and vapour phases in the column. The relevant variables to maintain this profile are the pressure, boil-up rate, reflux ratio and reflux temperature. Although composition control by direct analysis of the product streams is feasible, it is not practised due to the expense of metering equipment and because the response time is intolerable for accurate control purposes. Product samples should be analysed by chromatographic or other techniques, however, as a periodic check on quality.

Any change in vapour flowrate upwards through the column due to a variation in boil-up rate will alter the pressure at the top of the column and this provides a basis for control of the boil-up rate by regulation of reboiler steam pressure using a pressure controller as shown in Figure 13.4.

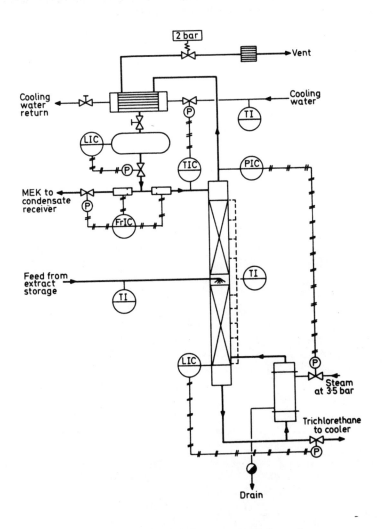

Figure 13.4 — *Control of solvent stripping column.*

The reflux rate will be controlled using measurement of liquid level in the accumulator (or reflux drum) to control total condensate flow and measurement and comparison of both distillate and reflux flowrates to ensure that the specified reflux ratio is maintained. Control of the reflux stream temperature will be achieved by the use of a temperature controller which actuates the cooling water inlet valve of the condenser.

Instrumentation of the main product distillation unit is similar to the stripping column and need not be discussed further. Figure 15.1 illustrates the control schemes devised in this Chapter together with other basic instrumentation but it should be emphasised that further investigations of the control required, using comprehensive operability studies for steady state and start-up conditions, should be completed before the proposed schemes are finalised.

Chapter 14

STORAGE OF PROCESS MATERIALS

14.1 General considerations

In all probability the MEK produced in this plant will be stored away from the process in a tank farm since it is most likely that all the product will be sold to other chemical and plastics manufacturers. Therefore it will be assumed that it is not necessary to provide any storage capacity for the MEK product; this being outside the scope of this design report. However the plant considered in this report will, most probably, be a part of a petro-chemical complex and will receive its secondary butanol feedstock from an adjacent plant. Consequently to prevent any discontinuity of operation of the MEK plant through production stoppages of the butanol process, a storage tank with two weeks capacity of 2-butanol should be sited conveniently to the MEK plant.

14.2 Secondary butanol storage tank

The exact location of this tank will depend on the layout of the whole petro-chemical complex and since two weeks supply would correspond to about 600m^3 capacity a vertical tank would be appropriate. Furthermore, from BSS 2654 (1973) a suitable tank would be one 10.0 m diameter, 10.0 m high and it would be constructed to conform with this British Standard. The design calculations will not be presented in this report and any reader interested is recommended to consult the above code.

Secondary butanol has a flash point of 24°C and is flammable. Therefore this tank must be fitted with a fixed roof and the appropriate safe venting valves. In addition the tank must also be fitted with a water line into the top and possess a foam fire extinguishing system. The tank must be located at least 6.5 m from any neighbouring tank and at least 30.0 m from the nearest building. The tank should be surrounded by a low concrete wall 1.5 m high and the tank should be located as far as is reasonably possible from the MEK plant.

14.3 Trichlorethane storage

Approximately 900 kg/h of 1.1.2 trichlorethane solvent will be circulated through the solvent extraction unit and small losses may be anticipated. In addition it may be anticipated that it will be necessary to empty the extraction column on occasion and therefore a storage tank with a minimum of 1.5 h capacity would be desirable. This will be a very small horizontal cylindrical storage tank and a tank 1.0 m diameter 2.0m long would be suitable, constructed to BS 2594 (1975) specifications. Trichlorethane will be contacted with water in the extraction unit and therefore there is the possibility of a small amount of hydrolysis occurring.

Consequently, this storage tank should be lined with a suitable plastic coating resistant to hydrochloric acid and insoluble in the trichlorethane. PTFE lining of a mild steel tank would be suitable since storage will be at ambient temperature.

1.1.2 trichlorethane is non-flammable but the vapour is narcotic and distinctly irritant to the membranes of the eyes, nose and throat. It is one of the more dangerous halogenated solvents and good ventilation must be ensured everywhere in the extraction plant and particularly the vents from this storage vessel must be directed into the stack gases after suitable washing to ensure that only trace concentrations are dispersed into the atmosphere.

Finally all the fire prevention directions issued in BS (CP) 3013, (1974) must be followed in all the storage vessels. These directions are too extensive to cite here and the reader is directed to this code of practice to ensure complete safety.

Chapter 15

FINAL CONSIDERATIONS

15.1 The final flow diagram

The plant proposed consists of a number of processing units, and the work of the chemical engineer has been essentially the determination of the size of the various equipment items. These items have been of varying complexity and their design has involved a knowledge of reaction rate in the case of the catalytic reactor, heat transfer in the case of the different ancillary heat exchangers, and simultaneous heat and mass transfer in the case of the design of the condenser, extraction and the distillation equipment. This work is the primary function of the chemical engineer but of equal importance is the arranging of the units as a whole and this is also the function of the chemical engineer. The arrangement is most easily accomplished after the final flow diagram has been prepared and this is shown in Figure 15.1*. The plant layout can then be arranged from this diagram as it should itemise everything involved in assembling the whole plant. Layout of the plant depends on the site available and it is not proposed to suggest a layout in this report.

15.2 Final remarks

It should not be thought that the work outlined in this report covers all the functions of the chemical engineer, or that sufficient detailed information has been evaluated and given for the plant to be built. If the plant were being erected, detailed placing of the units would have to be considered and the necessary schedules of pipes, electrical fittings, valves, *etc* prepared. The instrumentation suggested would require the provision of compressed air for its operation. Thus there is a great deal of planning work to be completed before it would be possible to form a picture of the proposed plant. The detailed design of the process vessels would be far more extensive than that given in Chapter 10. It would be the responsibility of a mechanical engineering department which would certainly exist and have its own special functions in any organisation which considered designing a plant of this complexity.

In conclusion, it should be stated that the present work is primarily an illustration of the method of setting out a design project but also it attempts to show how fundamental data of the literature and the laboratory can be translated and adapted to form the basis of a commercial project. A project report as prepared in a design department would be more concise than this but would deal with each plant item and the general plant layout in greater detail.

* *For Figure 15.1 see the fold-out sheet inside the back cover.*

REFERENCES

1. Gregory, S.A., Private Communication.
2. Austin, D.G., 1978 *Drawing Office Guide to Symbols used in the Petroleum, Chemical & Allied Industries,* (George Godwin, London).
3. Perona, J.J & Thodos, G., 1957 *AIChEJ,* 3: 230.
4. Thaller, L.H. & Thodos, G., 1960 *AIChEJ,* 6:369
5. Ford, F.E. & Perlmutter, D.D., 1964 *Chem Eng Sci,* 19:371.
6. Treybal, R.E., 1963 *Liquid Extraction* (McGraw Hill, New York).
7. Newman, N., Hayworth, C.B. & Treybal, R.E., 1949 *Ind & Eng Chem* 41:2039.
8. Jenson, V.G., & Jeffreys, G.V., 1977 *Mathematical Methods in Chemical Engineering,* 2nd Ed (Academic Press, London & New York).
9. Kolb, H.J., & Burrell, R.L., 1945 *J Am Chem Soc,* 67: 1084.
10. Coulson, J.M. & Richardson, J.F., 1977 *Chemical Engineering* (Pergamon Press, London).
11. Coulson, J.M., 1949 *Trans IChemE,* 27: 237.
12. Bird, R.B., Stewart, W.E. & Lightfoot, E.N., 1960 *Transport Phenomena* (Wiley, New York).
13. Hougen, O.A., & Watson, K.M., 1964 *Chemical Process Principles* (Wiley, New York).
14. Cooper, A.R. & Jeffreys, G.V., 1973 *Chemical Kinetics and Reactor Design* (Prentice Hall, New Jersey).
15. Rudd, D.F. & Watson, C.C., 1968 *Strategy of Process Engineering* (Wiley, New York).
16. "World Steel & Metal News" *Metal Bulletin* 6260 (January 1978)
17. "ECN Market Report" *European Chemical News,* (June 1978)
18. Holland, F.A., Moores, R.M., Watson, F.A. & Wilkinson, J.K. 1970, *Heat Transfer* (Heinemann, London).
19. Perry, J.H. 1972 *Chemical Engineers Handbook,* 5th Ed (McGraw Hill, New York).
20. Kern, D.Q. 1950 *Process Heat Transfer* (McGraw Hill, New York).
21. Butterworth, D., 1975 *Introduction to Heat Transfer* Engineering Design Guide 18 (BSI & CEI, London).
22. *Dimensions of Steel Pipe for the Petroleum Industry (Metric Units)* BS 1600 Part 2 1970 (BSI, London).
23. Frank, O. & Prichett, R.D., 1973 *Chemical Engineering,* 3:107.
24. Hughmark, G.A., 1961 *Chem Eng Prog,* 57:43
25. Scheiman, A.D., 1963 *Hydroc Proc* 42 (10): 165
26. Colburn, A.P. & Hougen, O.A., 1934 *Ind & Eng Chem,* 26: 1178.
27. McAdams, W.H., 1954 *Heat Transmission,* 3rd Ed (McGraw Hill, New York).
28. Friend, L. & Adler, G.D., 1958 *Transport Properties of Gases* (Northwestern Univ Press, Evanston),
29. Gilliland, E.R., 1934 *Ind & Eng Chem,* 25: 56.

30. Scheibel, E.G. & Othmer, D.F., 1944 *Trans AIChE*, 40: 611
31. Morris, G.A., & Jackson, J. 1953 *Absorption Towers*, (Butterworth, London).
32. Norman, W.S., 1961 *Absorption, Distillation & Cooling Towers* (Longmans, London),
33. Sherwood, T.K., & Holloway, F.A.L,; 1940 *Trans AIChE* 36: 319
34. Mumford, C.W. & Jeffreys, G.V., 1972 *J App Chem Biotechnol*, 22: 319.
35. Misek, T., 1963 *Coll Czech Chem Comm.*, 28: 426.
36. Meister, B.J. & Scheele, G.F., 1968 *AIChEJ*, 14:9.
37. Kolmogoroff, A.N., 1949 *Doklady Akad Nank, SSSR*, 66: 825.
38. Kung, E.Y. & Beckmann, R.B. 1961, *AIChEJ*, 7: 319.
39. Rose, P.M. & Kintner, R.C., 1966 *AIChEJ*, 12: 530.
40. Garner, F.H., Foord, T. & Tayeban, M. 1959., *J App Chem*, 9: 315.
41. Treybal, R.E., 1968 *Mass Transfer Operations*, 2nd Ed, (McGraw Hill, New York).
42. Amick, E.H., Weiss, M.A. & Kirshenbaum, M.S., 1951 *Ind & Eng Chem*, 43: 969.
43. *Unfired Fusion Welded Pressure Vessels*, BS 5500, 1976 (BSI, London).
44. Nelson, G.A., 1950 *Petroleum Refiner*, 29 (9): 104.
45. *Seamless and Welded Austenitic Stainless Steel Pipes and Tubes for Pressure Purposes*, BS 3605, 1973 (BSI, London).
46. *Tubular Heat Exchangers for General Purposes*, BS 3274, 1960 (BSI, London).
47. *Flanges and Bolting for Pipes, Valves and Fittings*, BS 4504, 1969 (BSI, London).
48. Bridgwater, A.V., 1977 *J Eff Water Treatment* 229 May.
49. Hasselbarth, J.E., 1977 *Modern Cost Engineering Techniques*, (McGraw Hill, New York).
50. Bridgwater, A.V. & Mumford, C.J., 1978 *Waste Recycling and Water Pollution Control Practice* (George Godwin, London).
51. Cran, J., 1977 *Eng & Process Economics*, 2: 13.
52. Bridgwater, A.V., & Timms, S.R., 1978 *Proc 5th Int Cost Eng Cong Utrecht*.
53. Taylor, J.H., 1977 *Eng & Process Economics*, 2: 259.
54. *Chemical Market Reporter*, June 1977.
55. Spiers, H.M., 1952 *Technical Data of Fuel*, 5th Ed, (British National Comm, London).
56. Peters, M.S. & Timmerhaus, K.D., 1968 *Plant Design and Economics for Chemical Engineers* (McGraw Hill, New York).
57. Clerk, J., 1964 *Chemical Engineering*, 71 (Oct 12): 232.
58. Bridgwater, A.V., Private Communication.
59. Elliott, D.M. & Owen, J.M., 1968 *The Chemical Engineer*, CE 377.
60. Chem Ind Safety & Health Council 1977 *A Guide to Hazard and Operability Studies*, (Chem Ind Association, London).
61. Wells, G.L., Seagrave, C.J. & Whiteway, R.M.C. 1977 *Flowsheeting for Safety* (IChemE, Rugby).
62. Kletz, T.A., 1978 *Chem & Ind*, 287, May.
63. Marsden, C., 1963 *Solvents Guide*, 2nd Ed, (Cleaver-Hulme, London).
64. Shinskey, F.G., 1967 *Process Control Systems*, (McGraw-Hill, New York).
65. Harriott, P., 1964 *Process Control*, (McGraw-Hill, New York).

Appendix A
VAPOUR HEAT CAPACITIES

Methods of prediction

Several methods of predicting heat capacity of vapours are available[1]. Among these are the Dobratz and the Group Contribution methods. For the purpose of comparison, both methods were used.

Dobratz method

Low pressure heat capacities of pure vapours and gases may be estimated with good accuracy with Dobratz's equation[1,2]

$$C_p^0 = 4R + n_r R/2 + \Sigma q_i C_{vi} + [(3n - 6 - n_r - \Sigma q_i)/\Sigma q_i]\Sigma q_i C_{\delta i}$$

where

- C_p^0 = low pressure heat capacity, cal/gmole K
- R = universal gas constant 1.987 cal/gmole K
- n_r = number of single bonds about which internal reaction of groups can take place
- q_i = number of bonds of the ith type
- n = number of atoms in the molecule
- Σq_i = total number of bonds in the molecule
- C_{vi} = Einstein function for stretching vibration for bonds of the ith type
- $C_{\delta i}$ = Einstein function for bending vibration for bonds of the ith type

Methyl ethyl ketone

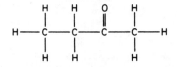

Figure A.1 — *Structure of the methyl ethyl ketone molecule.*

For the methyl ethyl ketone molecule, Figure A.1: $n_r = 3$; $n = 13$ and $\Sigma q_i = 12$. The molecule contains eight C–H bonds, three C–C bonds and one C = O bond. All these, coupled with the heat capacity constants of the heat capacity equation

$$C_p^0 = A + BT + CT^2$$

for each bond type as listed in Table A.1 below, lead to the derivation of an expression for the heat capacity of methyl ethyl ketone vapour as a function of temperature.

Table A.1 — *Heat capacity constants for methyl ethyl ketone.*

Bond	Stretching vibrations			Bending vibrations		
	A	$B \times 10^3$	$C \times 10^6$	A	$B \times 10^3$	$C \times 10^6$
C–H	−0.139	0.168	0.447	−0.579	3.741	−1.471
C–C	−0.339	3.564	−1.449	0.343	2.707	−1.150
C–O	−0.778	2.721	−0.759	−0.034	3.220	−1.341

$$\Sigma q_i C_{vi} = 8(-0.139 + 0.168 \times 10^{-3}T + 0.447 \times 10^{-6}T^2)$$
$$+ 3(-0.339 + 3.564 \times 10^{-3}T - 1.449 \times 10^{-6}T^2)$$
$$+ 1(-0.778 + 2.721 \times 10^{-3}T - 0.759 \times 10^{-6}T^2)$$
$$= -2.907 + 14.757 \times 10^{-3}T - 1.530 \times 10^{-6}T^2$$
$$\Sigma q_i C_{\delta i} = 8(-0.579 + 3.741 \times 10^{-3}T - 1.471 \times 10^{-6}T^2)$$
$$+ 3(0.343 + 2.707 \times 10^{-3}T - 1.150 \times 10^{-6}T^2)$$
$$+ 1(-0.034 + 3.220 \times 10^{-3}T - 1.341 \times 10^{-6}T^2)$$
$$= -3.637 + 41.269 \times 10^{-3}T - 16.559 \times 10^{-6}T^2$$
$$(3n - 6 - n_r - \Sigma q_i)/\Sigma q_i = [3(13) - 6 - 3 - 12]/12 = 1.5$$

Hence the heat capacity of methyl ethyl ketone vapour expressed as a function of temperature is

$$C_p^0 = 2.566 + 76.661 \times 10^{-3}T - 26.369 \times 10^{-6}T^2 \text{ (cal/g mole K)}$$
$$= 149 + 4.448T - 1.530 \times 10^{-3}T^2 \text{ (J/kg K)}$$

2-Butanol

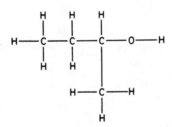

Figure A.2 — *Structure of the secondary butyl alcohol molecule.*

The molecule of 2-butanol, Figure A.2 contains nine C–H bonds, three C–C bonds, one C–O bond and one O–H bond. Thus for this molecule: $n_r = 4$; $n = 15$ and $\Sigma q_i = 12$. All these, together with the heat capacity constants of the heat capacity equation

$$C_p^0 = A + BT + CT^2$$

for each bond type as listed in Table A2 lead to the derivation of an expression for the heat capacity of 2-butanol vapour as a function of temperature.

Table A.2 – *Heat capacity constant for 2-butanol.*

Bond	Stretching vibrations			Bending vibrations		
	A	$B \times 10^3$	$C \times 10^6$	A	$B \times 10^3$	$C \times 10^6$
C–H	−0.139	0.168	0.447	−0.579	3.741	−1.471
C–C	−0.339	3.564	−1.449	0.343	2.707	−1.150
C–O	−0.458	3.722	−1.471	−0.665	3.757	−1.449
O–H	0.000	−0.240	0.560	−0.819	3.563	−1.267

Following a similar procedure as for methyl ethyl ketone, the heat capacity of 2-butanol vapour expressed as a function of temperature, T (K), is

$$C_p^0 = 0.697 + 89.351 \times 10^{-3} T - 30.343 \times 10^{-6} T^2 \text{ (cal/g mole K)}$$
$$= 39 + 5.043 T - 1.713 \times 10^{-3} T^2 \text{ (J/kg K)}$$

Group contribution method[1]

A certain molecule, of which methane is one, is first chosen as the base. Starting from this base molecule substitutions are made one after another so that the desired molecule is synthesised. The base molecule, as well as each of the substitutions, contributes a certain amount to the heat capacity of the molecule being synthesised and the vapour heat capacity of the desired molecule is obtained by summing up all these contributions.

Methyl ethyl ketone

Table A.3 – *Vapour heat capacity for methyl ethyl ketone.*

Total Group	Type	Contribution		
		A	$B \times 10^3$	$C \times 10^6$
CH_4	Base	3.42	17.85	−4.16
$CH_3.CH_3$	Primary	−2.04	24.00	−9.67
$CH_3.CH_2.CH_3$	1:1	−0.97	22.86	−8.75
$CH_3.CH_2.CH_2.CH_3$	1:2	1.11	18.47	−6.85
$CH_3.CH\,(CH_3)_2.CH_2.CH_3$	2:2	1.52	19.95	−8.57
$CH_3.C\,(CH_3)_2.CH_2.CH_3$	3:2	−1.19	28.77	−12.71
$CH_3.CO.CH_2CH_3$	=O	5.02	−66.08	30.21
	Total	6.87	65.82	−20.50

Table A.3 lists the results for the calculations for methyl ethyl ketone. Hence for methyl ethyl ketone vapour at T (K),

$$C_p^0 = 6.87 + 65.82 \times 10^{-3}T - 20.50 \times 10^{-6}T^2 \text{ (cal/g mole K)}$$
$$= 399 + 3.819T - 1.189 \times 10^{-3}T^2 \text{ (J/kg K)}$$

2-Butanol

Table A4 lists the results of the calculations for 2-butanol. At T (K), we have

$$C_p^0 = 6.21 + 88.27 \times 10^{-3}T - 32.41 \times 10^{-6}T^2 \text{ (cal/g mole K)}$$
$$= 350 + 4.982T - 1.829 \times 10^{-3}T^2 \text{ (J/kg K)}$$

Table A.4 — *Vapour heat capacity for 2-butanol.*

Total Group	Type	Contribution		
		A	$B \times 10^3$	$C \times 10^6$
CH_4	Base	3.42	17.85	−4.16
$CH_3.CH_3$	Primary	−2.04	24.00	−9.67
$CH_3.CH_2.CH_3$	1:1	−0.97	22.86	−8.75
$CH_3.CH_2.CH_2.CH_3$	1:2	1.11	18.47	−6.85
$CH_3.CH(CH_3).CH_2.CH_3$	2:2	1.52	19.95	−8.57
$CH_3.CH(OH).CH_2.CH_3$	−OH	3.17	−14.86	5.59
	Total	6.21	88.27	−32.41

Comments

Derivation of expressions of vapour heat capacities as a function of temperature for methyl ethyl ketone and 2-butanol by Dobratz and Group Contribution methods has led to the results shown in Table A5.

Table A.5 — *Heat capacity expressions.*

Compound	Heat capacity (J/kg K)	
	Dobratz	Group contribution
Methyl ethyl ketone	$1.49 + 4.448T - 1.530 \times 10^{-3}T^2$	$399 + 3.819T - 1.189 \times 10^{-3}T^2$
2-butanol	$39 + 5.043T - 1.713 \times 10^{-3}T^2$	$350 + 4.982T - 1.829 \times 10^{-3}T^2$

On inspection of this Table it is obvious that the two methods give distinctly different results. However, it is interesting to find that when values are substituted for T in the expressions (Table A6) their difference becomes unexpectedly insignificant for methyl ethyl ketone whereas the difference is rather considerable for 2-butanol. Another point worth mentioning is that the difference in the results using both methods gradually decreases with temperature and it may be deduced that the two methods will eventually converge at elevated temperatures since then the T and T^2 terms in the expressions become more and more predominant.

Table A.6 — *Comparison of prediction methods.*

Compound	Temperature (K)	Heat capacity (J/kg K)		Difference (J/kg K)
		Dobratz	Group Contribution	
Methyl ethyl ketone	300	1345.7	1437.7	92.0
	320	1415.7	1499.3	83.6
	340	1484.5	1560.0	75.5
	360	1552.0	1619.7	67.7
	380	1618.3	1678.5	60.2
	400	1683.4	1736.4	53.0
Secondary butyl alcohol	300	1397.7	1680.0	282.3
	320	1477.3	1757.0	279.7
	340	1552.5	1832.4	279.9
	360	1632.5	1906.5	274.0
	380	1708.0	1979.1	271.1
	400	1782.1	2050.2	268.1

In the present project, the working temperature ranges from 300 K to 400 K. Over this range, the two methods give quite different answers. Thus it is imperative to be able to select the right expression. It is not difficult to envisage that the Group Contribution method is less reliable since this method depends on the synthesis path chosen and the synthesis path may not be unique. Hence, it is decided that the expressions obtained from Dobratz's equation are the more reliable and these will be used throughout the whole of this design project.

Conclusion

The heat capacities for methyl ethyl ketone and 2-butanol vapour at a temperature of T (K) are predicted by Dobratz's equation to be:

and

$$C_p^0 = 149 + 4.448T - 1.530 \times 10^{-3}T^2 \text{ (J/kg K)}$$
$$C_p^0 = 39 + 5.043T - 1.713 \times 10^{-3}T^3 \text{ (J/kg K)}$$

respectively

Appendix B

HEAT OF REACTION

The heat of reaction was calculated from the standard heats of formation of the reacting species. These in turn were evaluated from heats of combustion.

for MEK $\qquad C_4H_8O + 5\tfrac{1}{2}O_2 \rightarrow 4H_2O + 4CO_2$

[3] Heat of formation $\tfrac{1}{2}H_2O = 34.19$ k cal/g mole
Heat of formation $CO_2 = 94.38$ k cal/g mole
Heat of combustion MEK = 582.3 k cal/g mole
∴ Heat of formation of MEK is given by:

$$\Delta H_f^{MEK} = (-4 \times 94.38) + (-8 \times 34.19) - (-582.3)$$
$$= -377.5 - 273.5 + 582.3$$
$$= -68.74 \text{ k cal/g mole}$$

for 2-butanol $\qquad C_4H_{10}O + 6O_2 \rightarrow 5H_2O + 4CO_2$

Heat of combustion of BUT = 635.91 k-cal/g-mole

$$\therefore \Delta H_f^{BUT} = (-4 \times 94.38) + (-10 \times 34.19) - (-635.91)$$
$$= -83.5 \text{ k cal/g mole}$$

and by definition $\Delta H_f{}^{H_2} = 0$

∴ Heat of Reaction at 25°C is thus

$$\Delta H_{298} = (\Delta H_f) \text{ products} - (\Delta H_f) \text{ reactants}$$
$$= (-68.7 + 0) - (-83.5)$$
$$= +14.8 \text{ k cal/g mole}$$

However, this must be corrected to the mean reaction temperature (723 K) using the expression presented by Cooper and Jeffreys[4]

$$\Delta H_{644} = \Delta H_{298} + \Delta\alpha(723-298) + \tfrac{1}{2}\Delta\beta(723^2 - 298^2) + \tfrac{1}{3}\Delta\gamma(723^3 - 298^3)$$

$$C_4H_{10}O \rightarrow C_4H_8O + O_2 \qquad \text{from Appendix A}$$

$\Delta\alpha = (2.57 + 6.95) - (-0.66) = 10.18$
$\Delta\beta = [(76.7 - 0.2) - (92.6)] \times 10^{-3} = -16.1 \times 10^{-3}$
$\Delta\gamma = [(0.48 - 26.41) - (-31.71)] \times 10^{-6} = 5.78 \times 10^{-6}$
$\Delta H_T = 14800 + [10.18(T-298)] - [8.05(T^2 - 88804) \times 10^{-3}]$
$\qquad\qquad + [5.78(T^3 - 26463592) \times 10^{-6})]$

∴ Heat of reaction at 723 K,

$$\Delta H_{723} = 14800 + (10.18 \times 425) - [80.5(52.2 - 8.9)]$$
$$+ [1.93(377 - 27)] \text{ cal/g mole}$$
$$= 14800 + 4326 - 3485 + 2031 = 17664 \text{ k cal/kg mole}$$
$$= 73900 \text{ kJ/kg mole}$$

Appendix C

EFFECTIVE THERMAL CONDUCTIVITY OF PACKED BED

Using the expression by Russell[5] which enables an approximate value of k_E to be made

$$\frac{k_E}{k_f} = \frac{\phi p^{2/3} + 1 - p^{2/3}}{\phi(p^{2/3} - p) + 1 - p^{2/3} + p} \text{ for } 0 < p < 1$$

where
$$p = \frac{\rho_{solid} - \rho_{bulk}}{\rho_{solid} - \rho_{vapour}}$$

and
$$\phi = \frac{k_{solid}}{k_f}$$

As the reactor tube is to be packed with 3.2 mm right circular cylinders

$$\rho_{solid} = 8500 \text{ kg/m}^3$$
$$^5k_{solid} = 125.4 \text{ W/m K}$$

ρ_{vapour} is assumed negligible for this calculation

Consider 100 m³ of the packed bed

Voidage, $\varepsilon = 0.393$

Volume occupied by cylinders = $(1 - \varepsilon)$ 100
$$= 0.607 \times 100$$

Mass of cylinders = $60.7 \times 8500 = 515\,950$ kg

∴ Bulk density, $\rho_{bulk} = 5159$ kg/m³

∴ $p = (8500 - 5159)/8500 = 0.394$ $p^{2/3} = 0.537$

The thermal conductivity of the reacting vapours was not available from the literature and consequently was predicted employing Eucken's method which has been modified by Smith[7] to predict the mean thermal conductivity of a hydrocarbon vapour mixture.

$$k_f = \sum_{i=1}^{n} \left\{ k_i \Big/ \left[1 + \sum_{j=1, j \neq i}^{n} \left(\frac{k_i^0}{k_{ij}^0}\right) \left(\frac{x_j}{x_i}\right) \right] \right\}$$

where
$$\left(\frac{k_i^0}{k_{ij}^0}\right) = \frac{1}{2\sqrt{2}} \left[1 + \left(\frac{k_i^0}{k_j^0}\right)^{1/2} \left(\frac{M_i}{M_j}\right)^{1/4} \right]^2 \left[1 + \frac{M_i}{M_j} \right]^{-1/2}$$

and where $k_i^0 = k_i/E'$ where E' is the Eucken type correlation factor which is given by

$$E' = (1 - \delta_f) + \tfrac{2}{5} \delta_f C_{pi}/R \text{ where } \delta_f = 0.885$$

A computer program was written to evaluate k_f for a given temperature, hence

$$k_f = 0.102956 \text{ W/m K at } 723 \text{ K}.$$

using equation

$$\frac{k_E}{k_f} = \frac{(\phi \times 0.537) + 0.463}{(\phi \times 0.143) + 0.857}$$

where $\phi = 1524/0.102956 = 1218.0$

Effective thermal conductivity, $k_E = 0.385$ W/m K

Appendix D
REACTOR DESIGN
LOGIC FLOW DIAGRAM

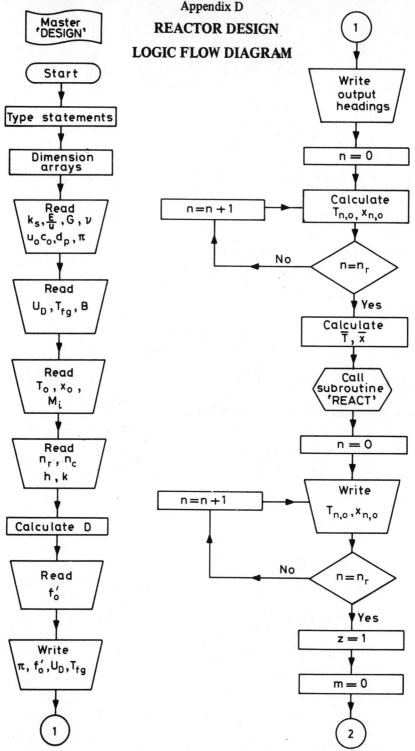

Figure D.1 – *Master 'DESIGN' section 1*   Figure D.2 – *Master 'DESIGN' section 2*

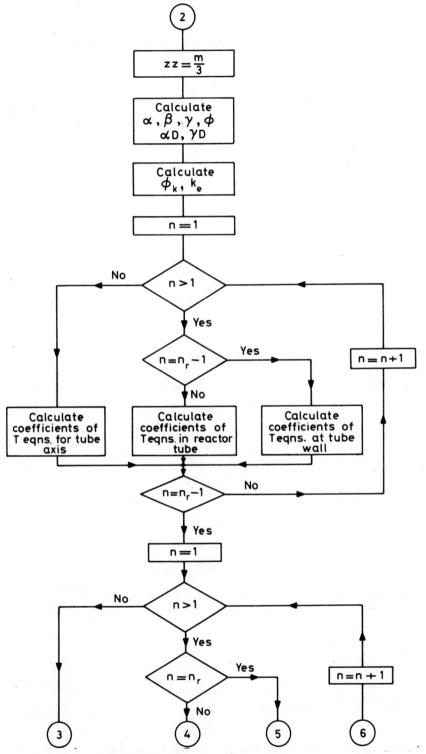

Figure D.3 – *Master 'DESIGN' section 3*

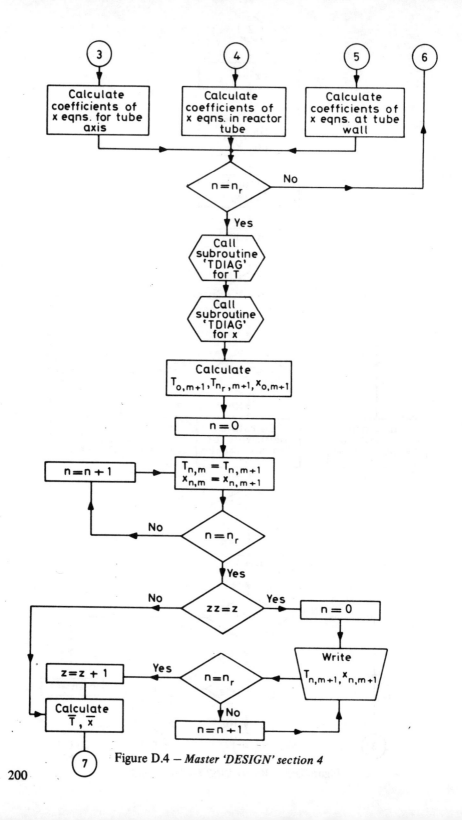

Figure D.4 – *Master 'DESIGN' section 4*

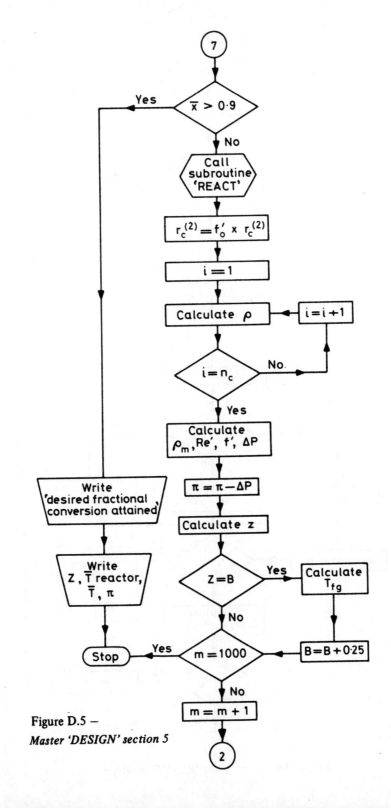

Figure D.5 —
Master 'DESIGN' section 5

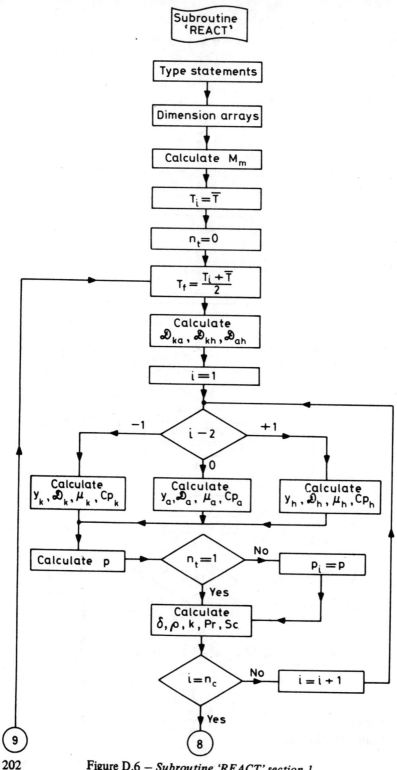

Figure D.6 – *Subroutine 'REACT' section 1*

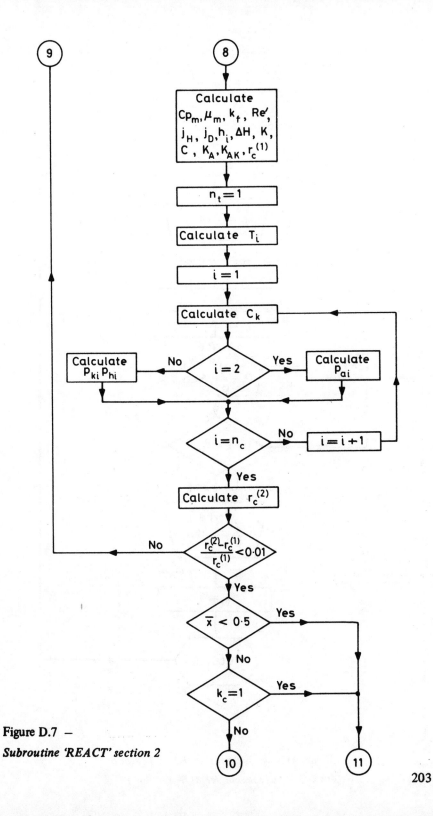

Figure D.7 —
Subroutine 'REACT' section 2

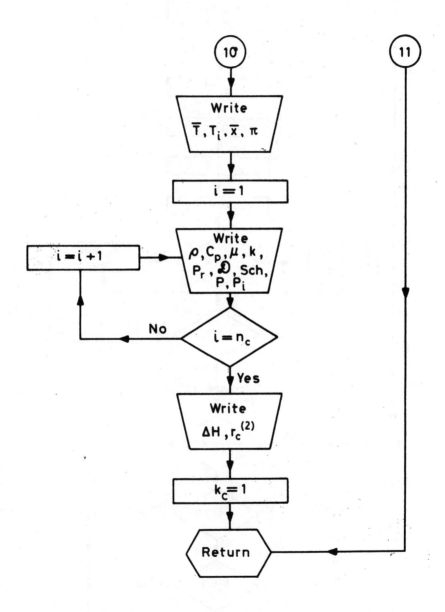

Figure D.8 — *Subroutine 'REACT' section 3*

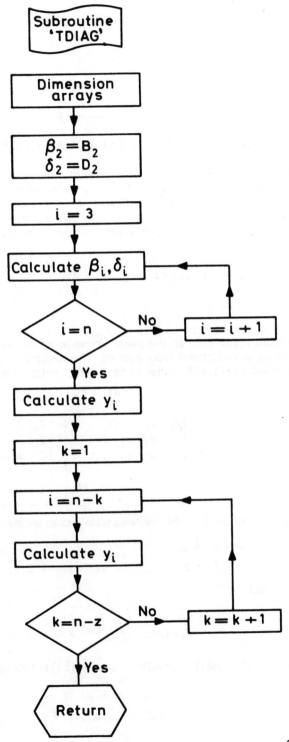

Figure D.9 — *Subroutine 'TDIAG'*

Appendix E

SOLUTION OF DIFFERENTIAL EQUATIONS FOR REACTOR DESIGN

The method which has been used to solve the finite difference equations derived in Section 4.5 is described by Lapidus[8]. The system of simultaneous linear equations for each length increment is treated using a modification of the Gaussian elimination technique known as the "Tridiagonal Method". Consider a system of $(n+1)$ simultaneous equations involving temperature with the form,

$$\left.\begin{array}{l} b_0 T_{0,m+1} + C_0 T_{1,m+1} = d_0 \\ a_r T_{r-1,m+1} + b_r T_{r,m+1} + c_r T_{r+1,m+1} = d_r \\ \qquad\qquad\qquad \text{where } r = 1, 2, 3, \ldots, n-1 \\ a_n T_{n-1,m+1} + b_n T_{n,m+1} = d_n \end{array}\right\} \quad \text{(E.1)}$$

Since d_r is a function of $T_{r-1,m}$, $T_{r,m}$ and $T_{r+1,m}$, the temperatures at the preceding length interval, this permits omission of the "$m+1$" subscript which will be taken as understood from now on. The coefficients a, b, c and d are known scalars and equation (E.1) may be expressed in matrix notation,

$$\mathbf{BT} = \mathbf{d}$$

or

$$\begin{bmatrix} b_0 & c_0 & & & \\ a_1 & b_1 & c_1 & & \\ & a_2 & b_2 & c_2 & \\ & & & \ddots & \\ & & & a_n & b_n \end{bmatrix} \begin{bmatrix} T_0 \\ T_1 \\ T_2 \\ \vdots \\ T_n \end{bmatrix} = \begin{bmatrix} d_0 \\ d_1 \\ d_2 \\ \vdots \\ d_n \end{bmatrix} \quad \text{(E.2)}$$

To solve equation (E.1) the following substitutions are made,

$$\beta_0 = b_0$$
$$\beta_r = b_r - (a_r c_{r-1}/\beta_{r-1}) \qquad r = 1, 2, \ldots, n \quad \text{(E3)}$$

and

$$\delta_0 = d_0$$
$$\delta_r = d_r - (a_r \delta_{r-1}/\beta_{r-1}) \qquad r = 1, 2, \ldots, n \quad \text{(E4)}$$

Equations (E3) and (E4) transform equation (E1) as follows,

$$T_n = \delta_n/\beta_n$$
$$T_r = (\delta_r - c_r T_{r+1})/\beta_r \quad \text{(E5)}$$

Therefore if ρ_r and δ_r are evaluated in the order of increasing r, it follows that equation (E5) can be used to calculate T (or x) in the order of decreasing r, that is,

$$T_n, T_{n-1}, \ldots, T_2, T_1, T_0.$$

This algorithm was incorporated into the computer program for the reactor design and the logic flow diagram is presented in Appendix D.

Appendix F

PHYSICAL PROPERTIES OF REACTOR FLUIDS

Calculation of effective diffusivity : linear velocity ratio

Using the correlation developed by Bernard & Wilhelm[9] involves the manipulation of a modified Peclet number, Pe

$$Pe' = \frac{d_p u}{E} = \left[\frac{(1.15/\lambda) - 0.15}{17.4(d_p/d_t)^{2.7} + 0.065}\right] Re'^{[4.28(d_p/d_t)^{1.8} - 0.05]}$$

where $\lambda = 1.0$ for spheres

tube diameter $d_t = 42$ mm $\quad \therefore d_p/d_t = 0.0756$

$\overline{Re'} = d_p G/\mu_m = 1000$

$$\frac{u}{E} = \left[\frac{314.9}{17.4(0.0756)^{2.7} + 0.065}\right] 1000^{[4.28(0.0756)^{1.8} - 0.05]}$$

$$= \left[\frac{314.9}{0.081311}\right] 1000^{-0.009}$$

$$= \frac{314.9 \times 0.939}{0.081311} = 3636 \text{ m}^{-1}$$

$\therefore$ Effective Diffusivity : Linear Velocity $(E/u) = 2.75 \times 10^{-4}$ m.

Calculation of surface area per unit volume for packed bed
Specific surface area,

$$S = \frac{\pi d_p^2 + \frac{1}{2}\pi d_p^2}{\frac{1}{4}\pi d_p^3} = \frac{6}{d_p} \text{ m}^2/\text{m}^3$$

for 3.2 mm diameter cylinders, $d_p = 25.4/8 = 3.175 \times 10^{-3}$ m

$$\therefore S = (6 \times 10^3)/3.175 = 1889 \text{ m}^2/\text{m}^3$$

assuming point contact for the random-packed bed surface area per unit volume, $v = s(1-\varepsilon) = 1889 \times 0.607 = 1147 \text{ m}^2/\text{m}^3$

Calculation of diffusion coefficients

For the purpose, Gilliland's correlation was used which is presented by Hougen and Watson[10]

$$D_{AB} = 0.0043 \frac{T^{3/2}}{\Pi(v_A^{1/3}+v_B^{1/3})^2} \sqrt{\left(\frac{1}{M_A}+\frac{1}{M_B}\right)}$$

where $\Pi = 2$ atm

v_A, v_B are molecular volumes which are found by summing the atomic volume contributions for each component[10].

hence
$$v_{MEK} = (4 \times 14.8)+(8 \times 3.7)+7.4 = 96.2$$
$$v_{BUT} = (4 \times 14.8)+(10 \times 3.7)+7.4 = 103.6$$
$$v_{H_2} = 14.3$$

Since there are three components diffusing from the catalyst/vapour interface and the bulk vapour phase, it is necessary to evaluate the mutual diffusivities of all components present. Also as the diffusivities are to be calculated along each increment of the reactor, they will be expressed as a function of absolute temperature for incorporation into the reaction rate subroutine

$$\therefore D_{MEK/BUT} = \frac{0.0043 T_K^{3/2}}{2(96.2^{1/3}+103.6^{1/3})^2} \sqrt{\left(\frac{1}{72}+\frac{1}{74}\right)} \text{ cm}^2/\text{s}$$

$$= \frac{0.0043 T_K^{3/2}}{2(4.58+4.69)^2} (0.0139+0.0135) \text{ cm}^2/\text{s}$$

$$= 6.855 \times 10^{-7} T_K^{3/2} \text{ cm}^2/\text{s}$$

$$\therefore D_{KA} = 0.6855 \times 10^{-10} T_K^{3/2} \text{ m}^2/\text{s}$$

similarly
$$D_{KH} = (22.50 \times 10^{-10}) T_K^{3/2} \text{ m}^2/\text{s}$$

and
$$D_{AH} = (21.80 \times 10^{-10}) T_K^{3/2} \text{ m}^2/\text{s}$$

Since the diffusing species do not constitute a binary system, it was necessary to weight the above coefficients on a mole fraction basis to avoid the necessity of evaluation of diffusivities for multicomponent situations which is an extremely complex procedure and not warranted for the situation. Therefore considering that a fractional conversion, x, has occurred, then

$$D_{MEK} = \left(\frac{1}{2x+1}\right) D_{KA} + \left(\frac{2x}{2x+1}\right) D_{KH}$$

$$D_{BUT} = 0.5(D_{KA}+D_{AH})$$

$$D_{H_2} = \left(\frac{1}{2x+1}\right) D_{AH} + \left(\frac{2x}{2x+1}\right) D_{KH}$$

These latter six equations were thus used in the reaction rate subroutine.

Evaluation of vapour densities

Initially, in order to account for non-ideality, it was considered desirable to use compressibility charts for the determination of molar volumes. However it was found that the compressibility factor, z, under the reaction conditions of temperature and pressure, was equal to unity for all components. Consequently, the ideal gas equation

$$pv = nRT.$$

was considered to be adequate and once again the densities were expressed in terms of temperature and partial pressure. The results of the correlations used were checked against known values to establish their applicability under the reaction conditions.

For one mole of gas, $n = 1$, $v = RT/p$

where $R = 0.08205$ litre atm/kg mole K

Therefore molar volume, $v = (82.050 \times T_K)/p$ cm^3/g mole

and specific volume $v/M = (82.05 \times T_K)/(10^{-3} \times pM)$ m^3/kg

where M is molecular weight

Therefore vapour density,
$$\rho_v = (pM \times 10^3)/(82.05 \times T_K) \text{ kg/m}^3$$

Evaluation of vapour viscosities

Method of Prediction: Vapour viscosities may be estimated by Arnold's Correlation[2]

$$\mu = \frac{27 \times 10^{-7} M^{1/2} T^{3/2}}{V_b^{2/3}(T + 1.47 T_b)}$$

where μ = viscosity of vapour (kg/m s)
  T = vapour temperature (K)
  T_b = boiling point of compound (K)
  V_b = molar volume of boiling liquid (cm^3/ gmole)
  M = molecular weight of compound

at 412°C
  μ_{H2} = 155.4 μP from Weast[3]
   = 1.554 × 10^{-4} P
   = 1.554 × 10^{-5} kg/m s

this compares well with μ_{H2} at 389°C = 1.53 × 10^{-5} kg/m from Perry[2].

Therefore using formula for effect of temperature in viscosity

$$\frac{\mu_2^0}{\mu_1^0} = \left(\frac{T_2}{T_1}\right)^{3/2} \frac{(T_1 + 1.47 T_b)}{(T_2 + 1.47 T_b)}$$

when $T_b = 20.3$ K for H_2

when $T_1 = 412°C = 685$ K

$$\mu_{H_2} = 1.554 \times 10^{-5} \left(\frac{T_2}{685}\right)^{3/2} \left[\frac{685+(1.47 \times 203)}{T_2+(1.47 \times 20.3)}\right]$$

$$= \frac{1.554 \times 10^{-5} \times 714.84}{(685)^{3/2}} \left(\frac{T_2^{3/2}}{T_2+29.8}\right)$$

$$= 6.196 \times 10^{-7} \left[\frac{T_2^{3/2}}{T_2+29.8}\right] \text{kg/m s}$$

of mean reaction temperature, *ie* 723 K.

$$\mu_{H_2} = 6.196 \times 10^{-7}[723^{3/2}/(723+29.8)] \text{ kg/m s} = 1.600 \times 10^{-5} \text{ kg/m s}$$

for the organic vapours, the correlation of Browning[11] produced acceptable values,

$$\mu = 1.286 M^{0.5} P_c^{0.667}(T/T_c) \times 10^{-7} \text{ kg/m s}$$

for MEK $T_c = 533$ K and $P_c = 43.3$ atm, therefore

$\mu_{MEK} = 1.286 \, (72)^{0.5} \, (43.3)^{0.667} \times (T_K/533) \times 10^{-7}$ kg/ms

$= 1.286 \times 8.5 \times 12.3 \times (T_K/533) \times 10^{-7}$ kg/m s

$= 2.522 \times 10^{-8} \, T_K$ kg/m s

similarly for 2-butanol

$$\mu_{BUT} = 2.735 \times 10^{-8} T_K \text{ kg/m s}$$

However, for heat transfer calculations, the mean vapour viscosity is required and was calculated from the expression given in Perry[2].

$$\mu_m = \frac{\Sigma y_i \mu_i (M_i)^{1/2}}{\Sigma y_i (M_i)^{1/2}} \quad \text{where } y_i = \frac{p_i}{p} = p_i$$

Therefore for a 50% fractional conversion at 723 K, mean vapour velocity,

$$\mu_m = \left[\frac{(0.6077 \times 8.5)+(0.6591 \times 8.6)+(0.5333 \times 1.415)}{0.33(8.5+8.6+1.415)}\right] \times 10^{-5} \text{ kg/m s}$$

$$= 1.877 \times 10^{-5} \text{ kg/m s}$$

Appendix G

SPECIFIC HEAT OF FLUE GAS

Firstly, it is necessary to estimate the composition of flue gas which will be assumed to be produced in a furnace burning either oil or hydrocarbon gases. The general equation for combustion of hydrocarbons is,

$$n(-CH_2-) + \tfrac{3}{2}nO_2 \rightarrow nCO_2 + nH_2O$$

If 8% excess oxygen is used, the number of moles of the constituents in the products of combustion were evaluated and from the mass fraction, the specific heat contributions at 775 K were added to calculate the mean specific heat of flue gas.

Table G.1 — *Data for specific heat of flue gas.*

Component	Moles	Mass	Mass Fraction	Specific heat (kJ/kgK)	Specific heat contribution (kJ/kgK)
O_2	$0.12n$	$3.84n$	0.013	1.088	0.014
N_2	$8.1n$	$226.8n$	0.775	1.130	0.876
CO_2	n	$44n$	0.150	1.172	0.176
H_2O	n	$18n$	0.062	2.092	0.129
Total	$10.22n$	292.64	1.000	—	1.195

∴ Specific Heat of Flue Gas at 775 K = 1.195 kJ/kg K

Appendix H

DATA FOR REBOILER DESIGN

Vapour density

Vapour density, $\quad \rho_v = \dfrac{pM}{0.08205 \times T_K} \text{ kg/m}^3$

where $p = 1$ atm, $M = 74$, $T_K = 373$ K.

$$\therefore \rho_v = \frac{74 \times 1}{0.08205 \times 373} \text{ kg/m}^3$$

$$= 2.418 \text{ kg/m}^3$$

Liquid density

$$\rho_L = 0.808 \times 1000 = 808 \text{ kg/m}^3 \text{ at 293 K from Weast}[3]$$

Correcting the density to 373 K using the expression from Perry[2]

$$\frac{(\rho_L - \rho_v)_{373}}{(\rho_L - \rho_v)_{293}} = \left(\frac{T_c - T_2}{T_c - T_1}\right)^{1/3}$$

for 2-butanol, $T_c = 538$ K and at 293 K $\rho_v = 3.078$ kg/m^3

$$\therefore \frac{\rho_L - 2.418}{808 - 3.078} = \left(\frac{538 - 373}{538 - 293}\right)^{1/3}$$

$\therefore$ Liquid Density, $\rho_L = 804.9 \ (0.674)^{1/3} + 2.418 \text{ kg/m}^3 = 709 \text{ kg/m}^3$

Mean specific heat

Now at 293 K, $C_{pL} = 2.26$ kJ/kg K[3] from Weast[3] correcting this value to 373 K.

$$C_{p100} = C_{p20}(w_1/w_2)^{2.8}$$

where $w = 0.1745 - 0.0838 \ (T/T_c)$ for $T_r \leqslant 0.65$ and $P \leqslant 10$ bar

now $T_c = 265°C$

$\therefore w_1 = 0.1745 - 0.0838 \ (20/265) = 0.168$

and $w_2 = 0.1745 - 0.0838 \ (100/265) = 0.1423$

$\therefore$ Liquid specific heat $C_{pL} = 2.26 \ (0.168/0.1423)^{2.8}$ kJ/kg K $= 3.60$ kJ/kg K.

For vapour,[12]

$$C_{pv} = f_1(a + bT) + e \cdot \exp(f_2/T)$$

where for 2-butanol

$a = 1.73$
$b = 8.20 \times 10^{-3}$

$$e = 1.774 \times 10^{-13}$$
$$f_1 = (1.0 + 1.444\, NG)$$
$$f_2 = (10220 + 454\, (N-H)\, G^2)$$
$$N = \text{number of carbon atoms}$$
$$H = \text{"degree of branching" (number of CH}_3 \text{ groups minus one)}$$
$$G = \text{"configuration factor"}\ (1.0 + 0.0022H)$$

at $212°F, N = 4; H = 1; G = 1.0022$.

$$C_{pv} = 6.8[1.73 + (8.2 \times 0.373)]$$
$$+ (1.774 \times 10^{-13})\exp\left\{\frac{10220 + [454 \times 3 \times (1.0022)^2]}{373}\right\}$$
$$= (6.8 \times 4.79) + (1.774 \times 10^{-13})\, e^{31.1}$$
$$= 37.9/74 = 0.51 \times 4.184 = 2.13\ \text{kJ/kg K}$$

$$\therefore \text{Mean Specific Heat},\ C_p = 2.86\ \text{kJ/kg K}$$

Surface tension

The surface tension of 2-butanol in contact with its own vapour at 353 K is[13] $\sigma_{353} = 17.4 \times 10^{-3}$ N/m. Correcting to the boiling temperature[2]

$$\frac{\sigma_{373}}{\sigma_{353}} = \left(\frac{T_c - T_2}{T_c - T_1}\right)^{1.2}$$

$$\sigma_{373} = 17.4\left(\frac{538 - 373}{538 - 353}\right)^{1.2} \times 10^{-3}\ \text{N/m}$$

$\therefore$ Surface Tension, $\sigma = 15.15 \times 10^{-3}$ N/m

Viscosity of Liquid

From Perry[2] $\mu = 0.275\, \rho_L^{1/2}$ now $\rho_L = 709/1000 = 0.709$, so viscosity $\mu = 0.275\, (0.709)^{1/2} = 0.231$ cP $= 0.231 \times 10^{-3}$ kg/m s

Thermal conductivity of fluid

From Perry[2] $k = 1.034\, C_p\, \rho_L^{4/3}/\alpha M^{1/3}$
where $\alpha = \lambda/T_b = 134.38 \times 74/373 \times 21 = 1.27$.
$\therefore$ Thermal Conductivity of Liquid,

$$k_r = \frac{1.034 \times 0.86 \times 0.709^{4/3}}{1.27 \times 74^{1/3}}\ \text{Btu/h ft °F}$$

$= 0.89 \times 0.63/1.27 \times 4.21 = 0.105$ Btu/h ft °F $= 0.1817$ W/m K.

Appendix I

VAPOUR THERMAL CONDUCTIVITIES

Method of Prediction

The approximation method of Eucken[2] was used to derive thermal conductivity expressions for hydrogen and for the vapours of methyl ethyl ketone and 2-butanol as functions of the absolute temperature (K). Eucken's approximation equation in SI units has the form:

$$k = 4186.6 \times \mu \times \left(\frac{C_p}{4186.8} + \frac{2.48}{M} \right)$$

where k = vapour thermal conductivity, W/m K
μ = vapour viscosity, kg/m s
C_p = heat capacity of vapour, J/kg K
M = molecular weight of vapour

Hydrogen

For hydrogen gas: $C_p = 12737 + 1.68T$ (J/kg K) from Perry[2]
$\mu = 3.0185 \times 10^{-8}\, T$ (kg/ms)
$M = 2.016$

$$k = 4186.6 \times 3.0185 \times 10^{-8} T \times \left(\frac{12737 + 1.68T}{4186.8} + \frac{2.48}{2.016} \right)$$

$$= 0.00054 + 5.07 \times 10^{-8} T^2 \text{ W/m K}$$

Methyl Ethyl Ketone

For methyl ethyl ketone vapour:

$C_p = 149 + 4.448T - 1.530 \times 10^{-3} T^2$ (J/kg K)
$\mu = 2.4269 \times 10^{-8}\, T$ (kg/ms)
$M = 72.10$

$$k = 1.016 \times 10^{-4} T (0.07 + 1.062 \times 10^{-3} T - 3.654 \times 10^{-7} T^2)$$
$$= 1.016 \times 10^{-4} (0.07T + 1.062 \times 10^{-3} T^2 - 3.654 \times 10^{-7} T^3) \text{ W/m K}$$

2-Butanol

For 2-butanol vapour:

$C_p = 39 + 5.043T - 1.713 \times 10^{-3} T^2$ (J/kg K)
$\mu = 2.4720 \times 10^{-8}\, T$ (kg/ms)

$M = 74.12$

$k = 1.035 \times 10^{-4} \, (0.043T + 1.204 \times 10^{-3} T - 4.091 \times 10^{-7} T^3) \, \text{W/m K}$

Thermal conductivity for vapour mixtures

According to Perry[2], the thermal conductivity of vapour mixtures may be most conveniently calculated from component conductivity values with the equation

$$k_m = \frac{\Sigma y_i k_i M_i^{1/3}}{\Sigma y_i M_i^{1/3}}$$

where k_m = thermal conductivity of vapour mixture, W/m K
y_i = mole fraction of component i
k_i = thermal conductivity of pure component i, W/m K
M_i = molecular weight of component i

Appendix J

PHYSICAL PROPERTIES OF THE CONDENSATE

General assumptions

An average condensate temperature of 45°C (318 K) is assumed and all the physical properties pertaining to the condensate are evaluated at this temperature. Furthermore, a mole fraction of 0.9 of methyl ethyl ketone in the condensate is also assumed.

Viscosity of condensate

At 45°C: viscosity of methyl ethyl ketone,[2] (liquid), $\mu_{MEK} = 0.38 \times 10^{-3}$ kg/m s
viscosity of 2-butanol (liquid), $\mu_{BUT} = 1.57 \times 10^{-3}$ kg/m s

Assuming a mole fraction of 0.9 of methyl ethyl ketone in the condensate and using the mixing rule,

$$\mu_m^{1/3} = x_1 \mu_1^{1/3} + x_2 \mu_2^{1/3}$$

viscosity of the condensate, $\mu_m = 4.53 \times 10^{-4}$ kg/m s

Density of condensate

At 20°C: density of methyl ethyl ketone (liquid) $\rho_{MEK} = 805$ kg/m³ density of 2-butanol (liquid), $\rho_{BUT} = 808$ kg/m³.

Since the critical constants for methyl ethyl ketone ($T_c = 535$ K; $P_c = 41.0$ atm)[13] and 2-butanol ($T_c = 536$ k; $P_c = 41.4$ atm)[13] are very similar, they can be considered as one entity with a hypothetical density at $20°C$ of 806 kg/m^3 and pseudocritical constants $T_c = 535$ K; $P_c = 41.1$ atm.

Using the Lu generalised chart[2] together with the relation,

$$\rho(45°C) = \rho(20°C)\frac{K(45°C)}{K(20°C)}$$

$k(20°C)$ is read for the T_r and P_r of the known density, ρ (20°C)

$T_r = 293/535;\ = 0.550\ \ P_r = 1/41.1 = 0.024$ and from Lu's chart, $K(20°C) = 1.040$.

$K(45°C)$ is read for the T_r and P_r of the desired density, ρ (45°C)

$T_r = 318/535 = 0.594;\ P_r = 1/41.1 = 0.024$ and from Lu's chart, $K(45°C) = 1.000$

Hence the density of the condensate is found to be 806 (1.000/1.040) = 775 kg/m^3.

Thermal conductivity of condensate

The thermal conductivity of each component in the condensate is first estimated by using Weber's original equation

$$k = 0.869\ C_p \rho^{4/3}/M^{1/3}$$

where k = thermal conductivity of component, (Btu/h ft °F)
           ρ = density of component, (g/cm^3)
           C_p = heat capacity of component, (Btu/lb °F)
           M = molecular weight of component

Methyl ethyl ketone

For methyl ethyl ketone: C_p = 0.549 Btu/lb °F
                                  ρ = 0.775 g/cm^3 (60°C)
                                  M = 72.10

Condensate thermal conductivity

Using the simple mixing rule,

$k_m = x_1\ k_1 + x_2\ k_2 = (0.9)\ (0.1412) + (0.1)\ (0.1860) = 0.1456$ W/m K

Appendix K
ECONOMIC ANALYSIS LOGIC FLOW DIAGRAM

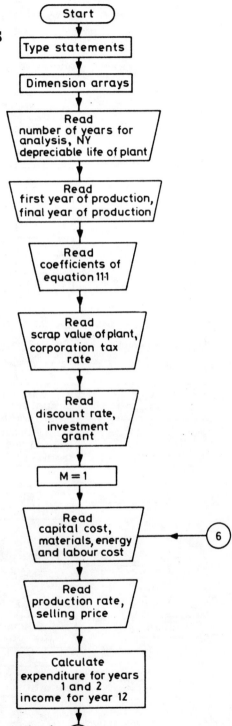

Figure K.1 —
Economic evaluation program section 1

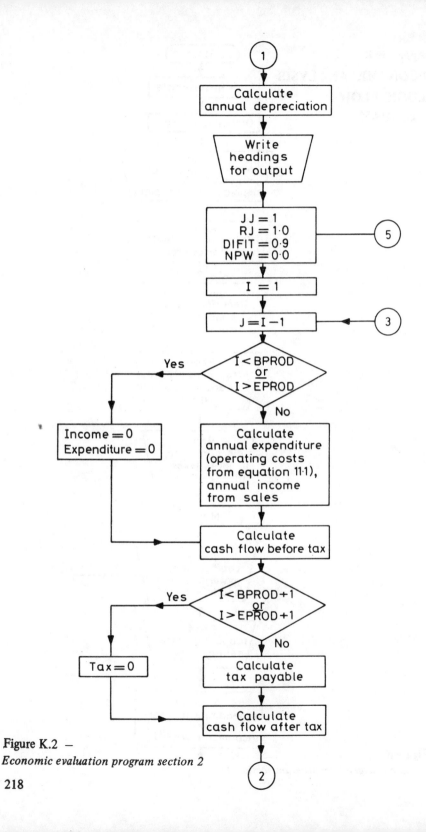

Figure K.2 —
Economic evaluation program section 2

218

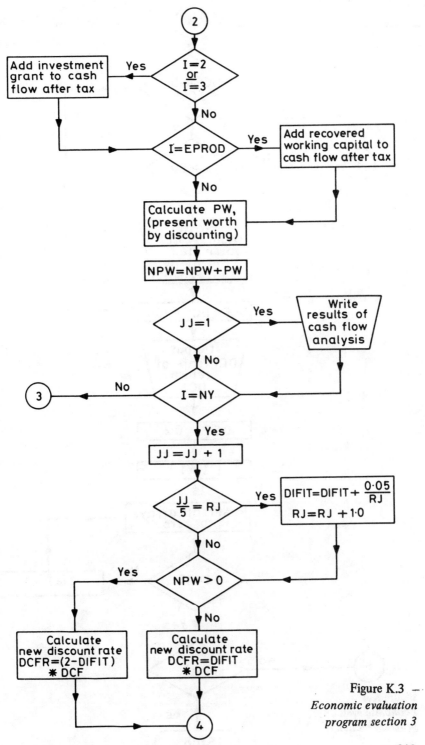

Figure K.3 — *Economic evaluation program section 3*

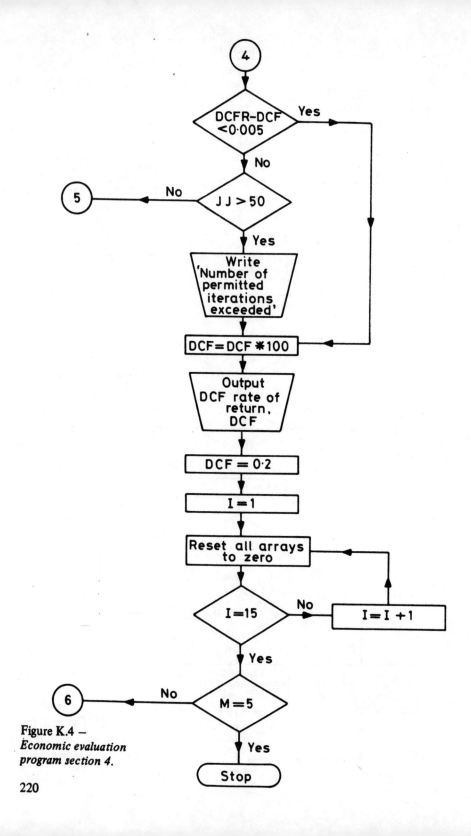

Figure K.4 — *Economic evaluation program section 4.*

APPENDIX REFERENCES

1. Reid, R.C. & Sherwood, T.K., 1958 *Properties of Gases and Liquids*, (McGraw-Hill, New York).
2. Perry, R.H. & Chilton, C.H. 1973. *Chemical Engineers' Handbook*, 5th Ed. (McGraw-Hill, Tokyo).
3. Weast, R.C., (Editor), 1977. *'Handbook of Chemistry and Physics*, 58th Ed. (Chemical Rubber Co., Press, Cleveland).
4. Cooper A.R. & Jeffreys G.V., 1973. *Chemical Kinetics and Reactor Design*. (Prentice Hall, New Jersey).
5. Russell, J. 1935. *J Am Ceramic Soc* 18: 1.
6. Vlack, V., 1968. *Elements of Materials Science*, 2nd Ed., (Addison-Wesley, London).
7. Smith, W.J.S., Durbin, L.D. Kobayashi, R., 1960. *J Chemical and Engineering Data*, 5 (3): 316.
8. Lapidus, L., 1962. *'Digital Computation for Chemical Engineers'* (McGraw-Hill, New York).
9. Bernard, R.A. & Wilhelm, R.H. 1950. *Chem Eng Prog* 46: 233.
10. Houghen, O.A. & Watson, K.M. 1964. *Chemical Process Principals – Part 3, Kinetics and Catalysis*, (Wiley, New York).
11. Browning, L.F., 1958. *Chemical Engineering*, 12: 169
12. Stromsoe, E. Renne, H.G. & Lydersen, A.L., 1970. *J Chemical and Engineering Data*, 15 (2); 286.
13. Young, T.F. & Harkins, W.D. 1928. *International Critical Tables*, 4: 451.